Value Addition & Quality Control of Meat Products: Challenges & Opportunities

Value Addition & Quality Control of Meat Products: Challenges & Opportunities

Editor

Prakash Shinde

scitus academics

Preface

Value addition in meat industry has always attracted the consumers and processors world over for the reason it gives variety, choice of products to suit their purse. Introduction of new varieties of meat product is constantly attempted by the processors and is also carefully watched by the consumers. Another aspect of value addition, which is gaining momentum, is the value added health foods. These products are designed to suit the nutritional requirement, diet restrictions of different categories of people. Energy foods for sports men, defense food and irradiated foods are developing with the specific purpose. Naturally when the value addition is done with the specific purpose the cost of the products takes the back seat while the quality with purpose is a priority like designer eggs, low fat meat recipes, high protein diets for children and self-heating cans etc. These are prepared as per the dietary guidelines and often have become an advertising tool for the manufacturing companies. While discussing about meat processing optimum utilization of by products is equally important for increasing profit and minimizing environmental pollution.

Editor

ABSTRACT

Poultry meat quality has been widely studied, and has become a growing demand of the international market. Parameters that affect meat quality are complex, and occur throughout the production chain. The constant concern with meat quality by the exporting sectors is a response to consumers' demands, and is achieved by increasing efficiency, and investments in personnel training on quality. Understanding where critical points are in the poultry meat production chain, and investing in solving critical problems may lead to better control and management, and consequent reduction of losses. Production and management practices, from farm to processing plant, play an important role in meat quality, and the use of technologies to reduce risk factors throughout the production chain will allow the production of better quality poultry meat not only for exports, but also for the domestic market. This review describes the main factors that influence poultry meat quality in the production chain.

INTRODUCTION

Brazil is the world's largest beef and poultry exporter, and fourth largest pork exporter, with total meat sales generating over USD 8 billion in 2005. Brazil achieved this position over the past decade, as Brazilian agriculture benefited from macroeconomic stability, high international commodity prices, currency devaluations, technological advancements, expansion in arable land area, and large capital inflows from domestic and direct foreign investment. Domestic policies (credit and tax-exemption programs) also spurred production and processors' incentives. Brazil's poultry meat exports account for 41% of global trade (UBA, 2005; Valdes, 2006).

Even though the United States is the largest poultry meat producer, Brazil is the second, and this position occasionally alternates. When adding to this figures other Latin American producing countries, such as Mexico, Colombia, and Argentina, the continent concentrates nearly 50% of the world's poultry meat production.

Broiler production in Brazil has become a very competitive activity during the last few years. Constant innovation lead to higher productivity while maintaining the same production cost, which is a desirable situation. One recent trend in Brazilian broiler production is to increase bird density, which requires the environmental control of poultry houses through the use of fan and fogging systems in order to achieve better live performance. Modern poultry production techniques, associated to better breeding management and health control, led to higher feed conversion ratio and improved production efficiency. Available literature (Xin *et al.*, 1994; Tinôco, 2001; Aradas *et al.*, 2005) indicates that improved environmental control of the broiler houses has played an important role in the increase in the productivity of the poultry industry, particularly in tropical areas, or those experiencing hot summer conditions.

According to Lima (2005), success in broiler production is directly associated to economic feasibility, qualified labor, intensive use of good practices, and a close link with the consumer market. Each of these items bears specific importance; however, they have to be considered in a multidisciplinary way, as these factors are overlapping. Average-size and small broiler production have met increasing difficulties to remain economically feasible in the present model of vertical structured production due to the necessary initial investment in housing, environmental control, etc.

Heier *et al.* (2002) recorded high mortality rates and other economical losses during the first weeks of broiler rearing, emphasizing that incorrect management and environmental control routines at this stage may cause significant losses.

Due to the market demand for quality, Brazilian processing plants are constantly implementing quality programs, as well as standards to obtain ISO 9000 certification. In addition, there is trend to produce broiler free from chemical, biological, and physical contaminants due to food safety concerns. This results in health and management constraints, as well as in limitations in the control of parasites and microorganisms inside poultry houses, and in bird and worker welfare (Lima, 2005).

PRODUCTION AND MANAGEMENT

Today, the market imposes several restrictions on poultry meat imports. The European Union, for instance, demands that birds need

Genetics

Breeding strategies have recently taken into account meat quality in order to meet consumers needs (Gaya, 2006). Different genetic attributes, such as meat quality and carcass yield, were studied by Moreira & Garcia (2003), who concluded that some commercial strains had better performance than others, when bird gender is considered.

According to Julian (2004), young broilers are very susceptible to variations in environmental temperature, which may lead to critical changes in metabolism due to unbalanced heat exchange. The author also points that some chemical drugs such as nicarbazin, and also chlorine and nitrophenols may make the problem yet more harmful by increasing body heat production.

Resistance to heat stress has become an important issue in the last years. Silva *et al.* (2001a) compared two broiler genetic strains: one with the gene (Na), which expresses low neck feathering, and one with normal neck feathering (nana), and observed that, when exposed to heat stress (38-42 °C) in an environmentally-controlled chamber, the birds with the Na gene presented better heat stress resistance as compared to the others. Studies report that reduced feathering volume may aid birds to optimize heat exchange in tropical conditions, becoming more tolerant to high temperature exposure (Marder & Arad, 1989; Macari & Gonzales, 1990; Singh *et al.*, 2001; Moreira *et al.*, 2006). In addition, non-nutritional factors, such as genetic inheritance and environment, play a vital role in the feathering rate, and may also contribute to poor feathering (Moreira *et al.*, 2006).

Nutrition

Garcia *et al.* (2005) studied the inclusion of sorghum replacing corn in broiler feeds, and found a significant negative correlation between meat pH decrease and corn replacement. Sorghum inclusion also affected in meat color, promoting paler meat.

The influence of vitamin E (as all-rac-a-tocopheryl acetate) supplementation on the sensorial quality of frozen chicken meat was studied. Two dietary treatments were compared as follows: *(1)* the control group received a diet containing respectively 30, 20, and 15 mg of a-tocopheryl acetate/kg of feed from 0-20, 21-38, and 38-

45 days of age; ***(2)*** the supplemented group received a starter diet containing 30 mg of á-tocopheryl acetate/kg of feed from 0 to 20 days of age, and a finisher diet containing 200 mg/kg of feed from 21 to 45 days of age. HPLC analyses showed that muscle á-tocopherol levels of the chickens fed the supplemented diet were 6-7-fold higher than those of the chickens fed the control diet. Meat quality was evaluated by sensorial as well as by instrumental techniques, such as (G-MS) thiobarbituric acid values (TBA) and gas chromatography-mass spectrometry. It was shown that vitamin E supplementation had a beneficial effect on sensorial data, and on oxidative stability of the meat, as measured by TBA. GC-MS analyses also showed that the concentration of aldehydes, which are considered responsible for rancid off-flavors, was much more important in the control samples as compared to the supplemented samples.

The effects of dietary a-lipoic acid (a-LA) on the pH value, AMP-activated protein kinase (AMPK) activation, and the activities of glycogen phosphorylase and pyruvate kinase in *postmortem* muscle were studied by Shen & Du (2005). According to the authors, dietary a-LA supplementation can suppress the activation of AMPK in*postmortem* muscle, down-regulate the activity of glycogen phosphorylase, and result in higher ultimate pH values in *postmortem* muscle. Therefore, dietary a-LA supplementation may be potentially reduce the incidence of pale, soft, and exudative (PSE) meat.

Health

In this text, the absence of health refer to biological abnormalities, which results in impaired fertility, hatchability, livability, productivity, or product quality. Many diseases cause mortality or drastic *postmortem*tissue alterations. The susceptibility or resistance of chickens and turkeys is partially determined by genetics. Few infectious agents are potent or virulent enough to overcome natural resistance in most birds. On the other hand, most infectious agents are possibly opportunistic. They have very little obvious effect on normal, healthy, well-cared-for birds, but may produce severe disease and drastic losses in birds stressed by rough handling, chilling during the first weeks, poor ventilation, etc.

Some of the most common origins of diseases in broiler flocks are described below as: ***1.*** Genetic-determined defects; **2.** Management related; ***3.*** Feed and water related; and, ***4.*** Infectious diseases.

Genetically Determined Defects: *Some Diseases Have an Evident Genetic Origin, and Many* others are suspected to have a genetic component. Anatomic deformities (for example, kinky back in broilers, wry neck, cystic or malformed kidneys, cystic right oviduct) are included in this category, as well as some behavioral (hysteria) and growth (poor feathering and tibial dyschondroplasia) problems. In general, major Brazilian breeding companies have done a good job in excluding these problems, and they seldom cause major constraints on productivity.

Management-Related Disease: *Unfortunately, Starvation, Dehydration, Cannibalism,* ammonia blindness, and a variety of other similar diseases are uncommon. These problems can be directly related to careless or inattentive management.

Feed- and Water-Related Diseases: *Diseases in This Category Include Nutritional* deficiencies, intoxications, and problems associated with chemical residues. The best approach for their prevention is monitoring feed and water in order to ensure the desired quality and to decrease the risk of hazardous contaminants.

Infectious Diseases: These Diseases Have So Many Possible Origins That It Is Difficult to specifically pinpoint them. The most common diseases are derived from the breeder flock. Any infectious agent that is transmitted from the breeder hen through the egg to the hatched chick or poult has its ultimate origin in the breeder flock. Mycoplasmosis, salmonellosis, and avian encephalomyelitis are among the diseases that can be transmitted from breeder flocks. At the hatchery, faulty sanitation can result in several diseases, such as omphalitis, salmonellosis, and aspergillosis.

Infectious agents, which are difficult to kill by routine cleaning and disinfection, are likely to be carried over from one flock to the next simply by survival in the contaminated premises. In this category, diseases caused by clostridial spores, the erysipela bacterium, staphylococci, infectious bursal agent, adenoviruses, Marek's disease, herpes-virus, etc are included.

Biological and mechanical carriers are particularly common sources of infectious diseases. Blood-sucking arthropods (mosquitoes, blackflies, mites, ticks, etc.) are known to be involved in the transmission of diseases, such as fowl pox and spirochetosis. Rodents and free-flying birds are dangerous sources of the causative agents of cholera and salmonellosis. Flies and beetles can be carriers of a number of infectious agents, and are known to be intermediate hosts of several internal parasites.

Tibial dyschondroplasia (TD) is a common leg abnormality that occurs in rapidly growing broilers, and it is well-established that growth rate reduction caused by the use of feed restriction decreases the incidence of leg abnormalities. However, feed restriction also results in lower body weights, and therefore it is not an interesting solution to leg problems for the commercial broiler producer. Several intermittent lighting regimes, as compared to continuous lighting, are effective in reducing leg abnormalities. In addition, intermittent light reduces mortality, while promoting similar or slightly better performance than continuous light. An experiment was conducted to determine if intermittent light could result in TD decreased without reducing body weight. Two lighting programs were tested in a total of 1,010 male and female broiler chicks from three genetic lines. The presence or absence of TD, and body weights were determined at four and seven weeks of age. Low intensity x-ray imaging was used for TD diagnosis (Lixiscope). Tibial dyschondroplasia was recorded as 0 for normal broilers or 1 for broilers with an abnormal mass of cartilage accumulation in the proximal tibiotarsus. Body weight gain and feed efficiency were calculated from 0 to 4 and from 4 to 7 weeks of age. The incidence of TD at four weeks of age was 15, 30, and 5% for the C, H, and L, lines, respectively. At seven weeks, incidence of TD varied from 47 to 11%, and was line-dependent. At four weeks, incidence varied from 21.4 to 12.0%, and at seven weeks, from 36.6 to 25.0% for males and females, respectively. Incidence of TD was reduced by intermittent light only in birds from a specific line. Tibial dyschondroplasia influenced body weights at four and seven weeks of age, independent from lighting program and sex. At four weeks of age, broilers with TD were 32 g heavier than broilers with no TD. At seven weeks of age, broilers with TD were 47 g lighter than those with no TD. It is possible that the effect of TD on body weights depends on broiler age, and on the severity of TD lesions. There were no differences in body weight or weight gain at four or seven weeks between the lighting programs. Body weights of seven-week-old birds submitted to constant or intermittent light were

2.31 and 2.32 g, respectively. Broilers under intermittent light had a better feed efficiency between zero and four weeks as compared to broilers exposed to constant light (1.67 and 1.71, respectively). Feed efficiency was similar from four to seven weeks in the two lighting programs (Maddocks, 2001).

Focal ulcerative dermatitis (FUD) is an ulcerative skin lesion over the breast area, usually on the unfeathered skin over the keel bone. Many factors, such as type of litter, ammonia concentration, feather loss, trauma, body weight, breast conformation, ammonia burns, and bacteria, were suggested as predisposing factors for both cellulitis and FUD, but the cause of breast lesions remains unknown (Vaillancourt & Barnes, 2003).

Cellulitis or infectious process has emerged as an economically significant disease syndrome of broiler chickens, resulting in increased condemnations and downgrading at processing, and costs associated with trimming and reprocessing affected carcasses. Cellulitis is characterized by a diffuse inflammatory reaction in the subcutaneous tissue. *Escherichia coli* (*E. coli*) is the predominant organism isolated from cellulitis lesions in a number of studies, being considered the causing agent of cellulitis. It was shown that skin trauma or damage by scratching increase cellulitis lesion formation by *E.coli*. It is believed that most cellulitis lesions observed at processing started during growout (Schrader *et al.*, 2004)

The impact of environmental risk factors on the incidence of cellulitis has not been evaluated in the United States yet. The causing agent, *E. coli*, is introduced through skin scratches during growout. Previous work suggested that the litter was an important reservoir for cellulitis-associated *E. coli*. Julian (2006) hypothesized that factors contributing to a positive environment for *E. coli* growth would increase the opportunity for the exposure of broilers to an *E. coli* infectious dose capable of initiating a cellulitis lesion. A prospective study of 304 flocks in five farms from two integrated broiler companies was conducted to determine the effect of environmental factors on the prevalence of cellulitis in California broiler flocks (Schrader *et al.*, 2004). Measured litter variables included *E. coli* and total gram-negative bacteria load (colony forming units/g dry matter), water activity, and pH. Cellulitis ranged from 0.2% to 6.0%. Significant associations were identified using linear regression among farm, litter moisture, days between flocks, and environmental temperature during the brooding period, gram-negative bacteria load in the litter during the brooding period, RH, grow-out, and *E. coli* load late in the growout period. The significant variation in cellulitis rate among farms, combined with a high association of litter moisture and days between flocks with cellulites, demonstrated that management decisions strongly influenced this disease syndrome. Based on these data and in previous findings, the authores advised managers to increase the number of days between flocks, and to perform a total clean out of the house if the previous flock had a high incidence of cellulitis at processing.

Harvesting and Transportation

There is evidence in the literature that catching often results in injury, especially when a large number of birds are caught with maximum haste by the catching team. Leandro *et al.* (2001) identified and quantified losses when comparing both manual and mechanized catching (16.5 to 7%, respectively). The authors found bruising in thighs, legs, and breasts of up to 25% of the harvested birds due to handling, catching, transportation, and unloading at the processing plant. However, most damage in the carcass was found during catching in the breast (11%), thighs (33%), and wings (38%).

Hip dislocation occurs when birds are caught in the broiler sheds and loaded into the transportation crates. Birds are usually held by one leg, and several birds are held in each worker hand. If one or more birds start flapping their wings, their hip twists, the femur detaches, and a subcutaneous hemorrhagic is produced, killing the bird. Dead birds with dislocated hip often have blood in the mouth, which was coughed up from the respiratory tract. Sometimes, this damage is caused by too much haste by the catchers (Gregory, 1998).

Debut *et al.* (2004) estimated the genetic variability between lines in terms of breast and thigh meat quality (pH drop, color, drip loss, and curing-cooking yield) by comparing a slow-growing French label-type line and a fast-growing standard line (SGL) of broilers exposed to different pre-slaughter stress conditions. Birds were slaughtered under optimal conditions or after exposure to 2h of transport or 2h of acute-heat stress (35°C). Relationships between meat quality and stress sensitivity were investigated by measuring struggle during shackling

Salmonella

Salmonella spp., Salmonella Enteritidis, and *Salmonella Typhimurium* are present in chicken processing plants, as evidenced in samples of feces, feathers, scalding water, taken during evisceration, chiller water, as well as samples of rinsing water of non-eviscerated, eviscerated, and chilled carcasses. Salmonella isolates were identified by multiplex-PCR assay using three sets of primers targeting the invA, pefA, and sefA gene sequences from *Salmonella spp., S. Typhimurium* and *S. Enteritidis,* respectively. *Salmonella spp.* was detected in 10% (29/288) of the samples, whereas serovars *Enteritidis* and *Typhimurium* were identified in 62% (7/288) of the samples. The results suggest the need to improve hygiene and sanitary standards in poultry slaughter lines, in addition to train food handlers, and to provide information to the consumers (Savell *et al.,* 2005; Cortez *et al.,* 2006).

Biochemical processes and structural changes that occur in muscle during the first 24 h *postmortem* play an important role in the ultimate quality and palatability of meat, and are influenced by the chilling processes to which carcasses are submitted after slaughter. Spray chilling, a system whereby chilled water is applied to carcasses during the early part of *postmortem* cooling, is used to control carcass shrinkage, and to improve chilling rates during evaporative cooling. Delayed chilling can be used to reduce or to prevent the negative effects of cold shrinkage; however, production constraints in high-volume facilities and food safety concerns make this method less useful in commercial settings. Electrical stimulation and alternative carcass suspension programs offer processors the opportunity to prevent most of all effects of cold shrinkage when traditional chilling systems are used.

FINAL REMARKS

The development of new and efficient processing systems, the adoption of advanced technologies, and the introduction of novel products that meet the market chain requirements and end-consumer needs have significantly contributed to increase global poultry meat consumption (Roenigk, 1998). Poultry production and processing technologies have become readily accessible, and are implemented on a worldwide basis, thereby allowing continuous expansion and competitiveness in this meat sector (Aho, 2001).

Processing and marketing of poultry ranges from live bird markets or very primitive on-site slaughter and sale, to highly sophisticated, fully automated and International Standards Organization (ISO) certified facilities and ready-to-eat convenience products in many parts of the world. Ironically, these extremes in processing and marketing methods are sometimes observed within the same country. The lack of inadequate refrigeration is probably the single largest obstacle to the marketing of many perishable foods, including poultry meat. In many developing countries, even minute improvements in cold storage capabilities have significantly increased trade, storage, distribution, and consumption of poultry products, whether domestic or imported (Babji, 2001). In addition, the adoption of modern freezing, packaging, and transportation technologies have provided large poultry companies the flexibility to export their high quality, mostly value-added premium cuts, to all corners of the world, especially at times of domestic surpluses and low market demands (Bilgili, 2006).

Consumers worldwide demand a protein supply that is safe, wholesome, nutritious, abundant, and affordable. However, food safety standards are currently not uniform, or equally enforced, around the world. Differences, for example, in hygiene or inspection regulations often lead to trade disputes and stagnation among the countries. Several international bodies, such as the Codex Alimentarius Commission of World Health Organization (WHO), the Food and Agricultural Organization (FAO), the International Commission on Microbiological Specifications for Foods (ICMSF), and the World Trade Organization (WTO) Agreements on Sanitary and Phytosanitary Measures (SPS) and Technical Barriers to Trade (TBT) develop and provide guidelines for the equivalency in international standards to protect the health of consumers and to ensure fair practices in food trade (Bilgili, 1999).

Welfare and good production practices on the farm, during transportation, and in the processing plant should be observed at all times, and be incorporated as part of an effective poultry production system. Adequate welfare procedures and proper hygiene must minimize dyschondroplasias, cellulitis, dermatitis, and exposure of carcasses to fecal material and potential pathogens, reducing the opportunities of cross contamination.

As the world moves towards a global market economy, poultry production will continue to increase in countries with important natural resources and that are economically competitive and politically open to business investments (Aho, 2001). Many countries or regions, on the other hand, will continue to struggle to compete in this market, either because of limited infrastructure (raw materials, land, utilities, labor, transportation) or political constraints (environmental and welfare concerns, tax structure, trade policies).

In 2003 and 2004, outbreaks of high pathogenic avian influenza in Asia led to the disruption of production and exports from key poultry producing nations. In Japan, the world's second largest poultry meat importing nation, poultry meat consumption significantly declined after these events. However, in the second half of 2004, a period in which US beef was excluded from Japan as a result of the discovery of BSE, limited supplies and subsequently higher beef prices resulted in a recovery in Japanese poultry meat consumption (Bailey, 2005).

Before poultry meat quality is addressed, it must be clearly defined how the term *quality* is related to poultry meat. This is a difficult task, because quality is "in the eye of the beholder". For example, someone trying to sell a product may consider quality in terms of how well the product is sold and how many people is willing to pay for it. However, this definition is incomplete, as it does not take into account product characteristics. As people only buy what they like, the consumer's perception of quality is the ultimate definition of quality, which needs to be met by poultry meat producing countries.

REFERENCES

1. Abu-Ruwaida AS, Sawaya WN, Dashti BH, Murad M, Al-Othman HA. Microbiological quality of broilers during processing in a modern commercial slaughterhouse in Kuwait. Journal of Food Protection 1994; 57:887892.
2. Agência Nacional de Vigilância Sanitária .Legislação. [Accessed dec. 08, 2006]. Available from: URL:http://www.anvisa.gov.br
3. Aho P. The new poultry elite. WATT Poultry USA, may, 20-24
4. Almeida Filho N, Rouquayrol MZ. Desenhos de pesquisa em epidemiologia: epidemiologia e saúde. 5. ed. Rio de Janeiro: MEDSI; 1999.

5. Almeida IAZC. *Salmonella*: sorotipos identificados na região de São José do Rio Preto/SP, no período de 1990-1999. Revista do Instituto Adolfo Lutz 2000; 59(1/2):33-37
6. AL-TECH, Comércio e Importação Ltda. Compreendendo e lidando com os efeitos das micotoxinas em rações e forragens para animais domésticos. [Acesso dez. 8,. 2006]. Disponível em: URL:http://www.altech.com.br/i01.htm
7. Aradas ME, Nääs I, Salgado IA. Comparing the thermal environment in broiler houses using two bird densities under tropical conditions. Agricultural Engineering International 2005; 8:1-9
8. Babji AS. Asian poultry processing adjusts to domestic market demands. World Poultry 2001; 17:22-24
9. Baley D. Beef and veal, pigs and poultry: meat outlook to 2009-10. Australian Commodities 2005; 12(1):65
10. Barbalho TCF, Almeida PF, Almeida RCC, Hofer E. Prevalence of *Listeria* spp. at a poultry processing plant in Brazil and a phage test for rapid confirmation of suspect colonies. Food Control 2005; 16:211216
11. Baruffaldi R, Oliveira MN. Fatores que condicionam a estabilidade de alimentos. In: ____. Fundamentos de tecnologia de alimentos. São Paulo: Atheneu; 1998. p.13-25
12. Bertol TM. Estresse pré-abate: conseqüências para a sobrevivência e a qualidade da carne em suínos. [Accessed out. 15, 2006]. Available from: URL: http://www.cnpsa.embrapa.br/?/artigos/2004/artigo-2004-n004.html
13. Bezerra V S. [Accessed dec. 08, 2006]. Available from: URL:http://www.embrapa.br/noticias/artigos/folder.2005-02-02.1550581232/artigo.2005-06-14.3711997257/.
14. Bilgili SF, Egbert WR, Hoffman DL. Research note: effect of postmortem ageing temperature on sarcomere lengh and tenderness of broiler Pectoralis major. Poultry Science 1989; 68(11):1588-1591
15. Bilgili SF. Assuring the safety of poultry products in a global market: The US perspective. In: Proceedings of the XIV European Symposium on the Quality of Poultry Meat; 1999; Bologna, Italy. p.513-519

16. Bilgili SF. Poultry Products and Processing. In: The International Market Place. Auburn: Department of Poultry Science Auburn University; 2006. p.8

17. Blokhuis HJ. Recent developments in european and international welfare regulations. World´s Poultry Science Journal 2004; 60(4):469-477

18. Bolis DA. Biosseguridade na criação alternativa de frangos. In: Conferência Apinco de Ciência e Tecnologia Avícola; 2001; Campinas, São Paulo. Brasil. p.223-234

19. Brasil. Ministério da Agricultura, Pecuária e do Abastecimento. Regulamento técnico da Inspeção tecnológica e higiênico-sanitária de carne de aves: Portaria DAS, nº. 210, de 10 de novembro de 1998, republicada em 05.03.99. [Accessed dec. 12, 2004]. Available from: URL:http://www.agricultura.gov.br/sda/dipoa/republport210.html

20. Brasil. Ministério da Agricultura, Pecuária e do Abastecimento. Regulamento técnico de métodos de insensibilização para o abate humanitário de animais de açougue: instrução normativa n.3, de 17 de janeiro de 2000. Brasília: [s.n]; 2000

21. Brasil. Portaria MAARA No.183 de 21 de março de 1996. Diário Oficial da União, Brasília (DF); 25 de março de 1996; Seção I. p. 4929

22. Bressan MC, Beraquet NJ. Efeito de fatores pré-abate sobre a qualidade da carne de peito de frango. Ciência Agrotécnica 2002; 26(5):1049-1059

23. Castellini C, Mugnai C, Dal Bosco A. Effect of conventional versus organic method of production on the broiler carcass and meat quality. Meat Science 2002; 60:219-225

24. Delazari I. Abate e processamento de carne de aves para garantia de qualidade. In: Conferência Apinco de Ciência e Tecnologia Avícola; 2001; Campinas, São Paulo. Brasil. p.191-203

25. Delazari I. Programas de segurança de alimentos na indústria da carne. Aveworld 2004; 1:44-49

26. Doll R, Peto R. The causes of cancer: quantitative estimates of avoidable risks of cancer in the United States today. Journal of National Cancer Institute 1981; 66:1191-308

27. Escudero-Gilete ML, González-Miret ML, Moreno Temprano R Heredia FJ. Application of a multivariate concentric method system for the location of *Listeria monocytogenes* in a poultry slaughterhouse. Food Control 2007; 18(1):69-75

28. FAO/WHO. Expert Committee on Food Aditives [JECFA]. Safety evaluation of certain food additives and contaminants – Aflatoxins. Geneva: World Health Organization; 1998

29. Fennema OR. Comparative water holding properties of various muscle food. Journal of Muscle Foods 1990; 1:363-381

30. Flatscher J, Willinger H. Bakterien-und pilzkeimzahlen in futtermitteln. wiener tierärztliche monatsschrift. In: Findlay, WPK. Fungi: folklore, fiction & fact. Surrey: Kingprint 1982; 68:282284

31. Fletcher DL. Poultry meat quality. World's Poultry Science Journal 2002; 58(2):131-145

32. Fletcher DL, Qiao M, Smith DP. The relationship of raw broiler breast meat color and pH to cooked meat color and pH. Poultry Science 2000; 79 (5):784-788

33. Garcia RG, Mendes AA, Costa C, Paz ICLA, Takahashi SE, Pelícia KP, Komiyama CM, Quinteiro RR. Desempenho e qualidade da carne de frangos de corte alimentados com diferentes níveis de sorgo em substituição ao milho. Arquivo Brasileiro de Medicina Veterinária e Zootecnia 2005; 57(5):634-643

34. Garcia RG, Mendes AA, Garcia EA, Nääs IA, Moreira J, Almeida ICL, Takita TS. Effect of stocking density and sex on feathering, body injury and breast meat quality of broiler chikens. Brazilian Journal of Poultry Science 2001; 1(4):1-9

35. Gasparetto KMPO. Sorotipos de *Salmonella* isolados de carcaças de frango e avaliação da susceptibilidade a antimicrobianos. Revista Ciência Farmacêutica 2001; 2(2):85-199

36. Gast RK. *Salmonella* infections. In: Saif YM editor. Diseases of poultry. 11th ed. Ames: Iowa State Press; 2003. p. 567-583

37. Gaya LG, Ferraz JBS. Aspectos genético-quantitativos da qualidade da carne em frangos. Ciência Rural 2006; 36(1):349-356

38. Groom GM. Factors affecting poultry meat quality. In: Options méditerranéennes Cambrigde (UK): Agricultural Development and Advisory Service (ADAS), Fisheries and Food, Cambridge (UK); 1990. (Série A-L'aviculture en Méditerranée

39. Heier BT, Hogasen HR, Jarp J. Factors associated with mortality in Norwegian broiler flocks. Preventive Veterinary Medicine 2002; 53(1-2):147-158

40. Julian RJ. Production and growth related disorders and other metabolic diseases of poultry: a review. The Veterinary Journal (in press). [Accessed dec. 10, 2006]. Available from: URL:http://www.sciencedirect.com

41. Kranen RW. Haemorrhages in muscles of broiler chickens. World´s Poultry Science Journal 2000; 56(2):93-126

42. Leandro NSM. Efeito do tipo de captura dos frangos de corte sobre a qualidade de carcaça. Ciência Animal Brasileira 2001; jul-dez: 97-100

43. Leslie JF, Pearson CHAS, Nelson PE, Toussoun TA. *Fusarium* spp. from corn, sorghum and Soybean Fields in the Central and Eastern United States. Phytopathology 1987; 80:334350

44. Lima AMC. Avaliação de dois sistemas de produção de frango de corte: uma visão multidisciplinar [tese].Campinas (SP): Universidade Estadual de Campinas; 2005

45. Maciorowski KG. Incidence, sources, and control of food-borne Salmonella spp. In poultry feeds. World´s Poultry Science Journal 2004; 60(4):446-457

46. Maddocks SA. Behavioural and physiological effects of absence of ultraviolet wavelengths for domestic chicks. Animal Behaviour 2001; 62:1013-1019

47. Matheus DP, Rudge AC, Gomes SMM. Ocorrência de *Salmonella* spp em carne de frango comercializada no município de Bauru (SP), Brasil. Revista do Instituto Adolfo Lutz 2003; 62(2):111-115

48. Mclnerney J. Animal welfare, economics and policy: report on a study undertaken for the farm & animal health economics. Londres: Division of DEFRA; 2004

49. Mendes AA. Rendimento e qualidade da carcaça de frangos de corte. In: Conferência Apinco de Ciência e Tecnologia Avícolas; 2001; Campinas, São Paulo. Brasil. p.79-99

50. Miragliotta MY. Avaliação de dois sistemas de ventilação de densidade e sua influência sobre níveis de amônia em sistema de produção de frangos de corte [dissertação]. Campinas (SP): Universidade Estadual de Campinas; 2001.

51. Moreira J, Garcia RG. Qualidade da carne de peito de frango de corte. Revista Nacional da Carne jul/2003; 317:14-19
52. Moreira J, Mendes AA, Garcia RG, Garcia EA, Roça RO, Nääs IA, Dalanezi JA, Pelícia K. Evaluation of strain, dietary energy level and stocking density on broiler feathering. Revista Brasileira de Ciência Avícola 2006; 8(1):15-22
53. Moss MO. Fungal Biotechnology Roundup. Mycologist 1990; 21:55-58
54. Northcutt JK. Factors affecting poultry meat quality. Bulletin 1157 1997. [Accessed nov. 18, 2006]. Available from: URL:http//pubs. caes.uga.edu
55. Oliveira CM, Carvalho DI. Rendimento e lesões em carcaças de frangos de corte criados em diferentes camas e densidades populacionais. Ciência Agrotécnica 2002; 26(5):1076-1081
56. Quevedo A. Frango à brasileira. [Accessed aug. 16, 2005]. Available from: URL:http://www.aviculturaindustrial.com.br/site/dinamica.asp?id=12036&tipo_tabela= negocios&categoria= mercado_interno
57. Roenigk WP. Poultry will overtake pig meat consumption. World Poultry 1998; 14(12):14-16
58. Santos DMS. *Salmonella* em carcaças de frango congeladas. Pesquisa Veterinária Brasileira 2000; 20(1):39-42
59. Savenije B, Lambooij E, Gerritzen MA, Venema K, Korf J. Effects of Feed Deprivation and Transport on Preslaughter Blood Metabolites, Early Postmortem Muscle Metabolites, and Meat Quality. Poultry Science 2002; 81:699708
60. Schrader JS, Singer RS, Atwill ER. A prospective study of management and litter variables associated with cellulitis in California broiler flocks. Avian Diseases 2004; 48:522-530
61. Silva MAN, Silva IJO, Piedade SMS. Resistance to heat stress of naked neck broilers. Revista Brasileira de Ciência Avícola 2001a; 3(1):27-33
62. Silva N, Junqueira VCA, Silveira NFA. Manual de métodos de análise microbiológica de alimentos. 2. ed. São Paulo: Varela; 2001b
63. Soerensen P, Su G, Kestin SC. Effects of age and stocking density on leg weakness in broiler chickens. Poultry Science 2000; 79:864870

64. Tinôco IFF. Avicultura Industrial: novos conceitos de materiais, concepções e técnicas construtivas disponíveis para galpões avícolas Brasileiros. Revista Brasileira de Avicultura 2001; 3 (1):01-26

65. UBA: União Brasileira de Avicultura. Relatório anual 2003/2004. [Accessed feb. 20, 2005]. Available from: URL:http://www.uba.br/2005

66. Valdes C. Brazil Emerges as Major Force in Global Meat Markets. Livestock, Dairy, and Poultry Outlook, Mildred Haley. LDP-M-137 – LDP-M-139, USDA, Economic Research Service, January 2006. [Accessed dec. 08, 2006]. Available from: URL: www.ers.usda.gov/publications/ldp/.]

67. Warriss PD, Knowles TG, Brown SN, Edwards JE, Ketlewell PJ, Mitchell MA, Baxter CA. Effects of lairage time on body temperature and glycogen reserves of broiler chickens held in transport modules. Veterinary Record 1999; 145:218222.

Chapter 2

Antimicrobial Edible Films and Coatings for Meat and Meat Products Preservation

Irais Sánchez-Ortega[1,2], Blanca E. García-Almendárez[1], Eva María Santos-López[2], Aldo Amaro-Reyes[1], J. Eleazar Barboza-Corona[3], and Carlos Regalado[1]

[1]DIPA, PROPAC, Facultad de Química, Universidad Autónoma de Querétaro, 76010 Querétaro, QRO, Mexico

[2]Área Académica de Química, Instituto de Ciencias Básicas e Ingeniería, Universidad Autónoma Del Estado de Hidalgo, Ciudad Del Conocimiento, Carr. Pachuca-Tulancingo Km 4.5 Col Carboneras, 42184 Mineral de la Reforma, HGO, Mexico

[3]División Ciencias de la Vida, Universidad de Guanajuato, Campus Irapuato-Salamanca, 36500 Irapuato, GTO, Mexico

ABSTRACT

Animal origin foods are widely distributed and consumed around the world due to their high nutrients availability but may also provide a suitable environment for growth of pathogenic and spoilage microorganisms. Nowadays consumers demand high quality food with an extended shelf life without chemical additives. Edible films and coatings (EFC) added with natural antimicrobials are a promising preservation technology for raw and processed meats because they provide good barrier against spoilage and pathogenic microorganisms. This review gathers updated research reported over the last ten years related to antimicrobial EFC applied to meat and meat products. In addition, the films gas barrier properties contribute to extended shelf life because physicochemical changes, such as color, texture, and moisture, may be significantly minimized. The effectiveness showed by different types of antimicrobial EFC depends on meat source, polymer used, film barrier properties, target microorganism, antimicrobial substance properties, and storage conditions. The perspective of this technology includes tailoring of coating procedures to meet industry requirements and shelf life increase of meat and meat products to ensure quality and safety without changes in sensory characteristics.

INTRODUCTION

Animal origin foods (AOF) constitute a good nutrients source for human diet, where their protein provides high biological value and essential amino acids which complement the quality of cereals and other vegetable proteins [1]. However, AOF are susceptible to chemical deterioration and microbiological spoilage and therefore represent a high risk for consumer health, in addition to producer economic losses. According to the Centers for Disease Control (CDC), every year foodborne illnesses account for about 48 million cases, 3,000 deaths, and 128,000 hospitalizations, reaching US $77.7 billion economic burden in the United States. In addition, reduced consumer confidence, recall losses, or litigation costs should be met by the food industry, whereas public health agencies pay the cost of responding to illnesses and outbreaks [2]. Losses can be greater in countries where less stringent regulation system and sanitary control is practiced.

Outbreaks involving AOF comprise 40% of total US reported cases [3]. The presence of foodborne pathogens in a country food supply not only affects the health of local population but also represents a potential for pathogens spread by tourists and consumers where these food products are exported [4].

Edible coatings are food grade suspensions which may be delivered by spraying, spreading, or dipping, which upon drying form a clear thin layer over the food surface. Coatings are a particular form of films directly applied to the surface of materials and are regarded as part of the final product [5]. On the other hand, edible films are obtained from food grade phylogenic suspensions that are usually cast over an inert surface, which after drying can be placed in contact with food surfaces. Films can form pouches, wraps, capsules, bags, or casings through further processing and one of the main differences between films and coatings is their thickness.

The use of films in foods dates back to the 12th century in China where waxes were used to coat citric fruits to retard water loss, whereas the first edible film used for food preservation was made in the 15th century from soymilk (Yuba) in Japan. In England lard or fats were used as coating to prolong shelf life of meat products in the 16th century and in Europe; this process was known as "larding" [6, 7]. In the nineteenth century, a US patent was issued in relation to preservation of meat products by gelatin coatings [7, 8].

Edible films and coatings (EFC) are an alternative to extend the shelf life of AOF by acting as barriers to water vapor, oxygen, and carbon dioxide and as carriers of substances to inhibit pathogenic and spoilage microorganisms. Natural antimicrobial agents may be incorporated into the corresponding suspensions, adding functionality to edible films and coatings, leading to the antimicrobial edible films and coatings (AEFC) obtaining.

There is increased interest in development and use of AEFC to preserve meat quality for longer shelf life periods while maintaining food safety, which is based on consumers demand for natural and safe products. Industry is concerned about these issues, while keeping competitive production costs [9]. Other key issues are sustainability through the use of biodegradable packaging materials and applications of by-products from the food industry that can generate added value [5].

Pork sausages	Chitosan films	Green tea extract 20% (w/v) in the chitosan film-forming solution	Mesophilic bacteria, yeasts and molds, lactic acid bacteria (LAB), not inoculated	No inoculation	4°C, 20 d Sausages wrapped with films, packaged into a pouch of low density polyethylene coated with polyamide plastic bag and heat-sealed	On day 12, faster growth in control samples for total viable count and molds and yeast was found; no difference for LAB	[42]
Cooked cured ham	Polylactic acid (PLA) films	Lauric arginate (LAE) (0% to 2.6%, w/w)	Listeria monocytogenesand Salmonella enterica serovar typhimurium (105 CFU/mL)	Inoculum of bothL. monocytogenesand S. typhimuriumonto surface of the sliced ham	4°C, 7 d Slices sterilized with UV on each side prior to inoculation Inoculated samples wrapped with LAE-coated PLA film and stored in closed plates	LAE-coated PLA film (2.6%) showed a significantly greater antibacterial activity, with L. monocytogenesand S.typhimurium levels reduced to <2 log CFU/film after 24 h exposure and remaining at this low level for the next 6 d	[43]

Table 2: Use of antimicrobial films and coatings in poultry

Product	Coating material	Antimicrobial compound	Target microorganism	Inoculation technique	Conditions	Results	Ref.
Chicken breast (ready-to-eat cooked chicken cubes)	Zein coatings dissolved in propylene glycol (ZP) or ethanol (ZE)	Nisin (N) (1000 IU/g) and/or calcium (CP) propionate (1% w/v)	L. monocytogenesV7 (low and high inoculum 2.67 log CFU/g and 6.89 log CFU/g, respectively, at 4°C)	Cubes immersed in 24 h broth cultures for 30 s, allowed to drip free of excess inoculum, and dried. Frozen samples were irradiated (3.0 kGy) and kept frozen until used	4°C or 8 °C, 24 d. Cubes boiled in water bath for 20 min, were inoculated, then dried, followed by dipping in edible ZP or ZE, with and without antimicrobials. Air dried samples (20 min) stored in sterile bags	L. monocytogenes was reduced by 4.5–5 log CFU/g relative to the control after high dose and 16 d at 4°C, with more significant effect when N was added to the films. Low inoculum dose and using ZPNCP film caused complete inhibition from 4 to 24 d, either at 4° or at 8°C	[18]

Turkey frankfurter	WPI coatings	Grape seed extract (GSE, 1.0–3.0% w/v), nisin (N, 6–18 kIU/g), malic acid (MA 1.0–3.0%; w/v), EDTA (1.6 mg/mL), and their combinations	L. monocytogenes, E. coli O157:H7, and Salmonella typhimurium(106 CFU/g)	Samples were defrosted and dipped into cultures of 106 CFU/mL of L. monocytogenes, E. coli O157:H7, or S. typhimurium for 1 min at room temperature. Inoculated samples were then air dried under laminar flow conditions	4°C, 28 d Samples were dipped in film-forming solutions (1 min) and dried (10 min, room temperature). Samples were then packed individually in sterile bags, and stored	L. monocytogenes decreased to 2.3 log/g (N, 6000 IU/g; GSE 0.5%; MA 1.0%). S. typhimurium decreased to 5 log CFU/g using any antimicrobial, whereas E. coli decreased to 4.6 log cycles using N, MA and EDTA. All reductions were relative to the control	[53]

Chicken breast	High methoxyl pectin 11400 with apple puree films	Carvacrol (C) or cinnamaldehyde (CM) at 0.5–3% (w/w).	Salmonella enterica serovarEnteritidis or E. coli O157:H7 (ATCC 35150) (10^7 CFU/g)	Inoculum was dispersed on the surface as droplets	23°C or 4°C, 3 d Samples were dipped in boiling water (40 s), plated and exposed in a bio-hood for drying (30 min). Sample was flipped over and inoculated in a similar way. Meat was wrapped using appropriate edible films	At 23 °C, films with 3% antimicrobials showed the highest reductions (4.3–6.8 log CFU/g) of both S.Enteritidis and E. coliO157:H7. At 4°C, C exhibited greater activity than CM. Relative to control samples, films with 0.5–3% C reduced S.Enteritidis by 1.6–3 log CFU/g, whereas 1–3% CM films reduced its population by 1.2–2.8 log CFU/g. Films with 0.5–3% C reduced 1–3 log E. coli whereas 1–3% CM films inhibited 0.2–1.2 log CFU. Treatments were at 4°C	[55]

Chicken breast fillets	Chitosan (CH) coating, deacetylation degree of 75–85%	Chitosan (1.5 % w/v) and/ or oregano oil 0.25% v/w (OO)	Mesophilic microorganisms, Pseudomonasspp. and Brochothrix thermosphacta	No inoculation	12, 18 y 21 d, 4°C Samples were dipped into the chitosan solution (1.5 min) and drained. Sterile OO was added to the surface. Fillets were packed in plastic pouches, and stored in a modified atmosphere	Shelf life of chicken fillets can be extended using either OO and/or CH, by approximately 6–21 d	[49]
Chicken breasts fillets	Chitosan (CH) films	CH or CH-LAE (1, 5 or 10%,by weight)	Mesophiles, psychrophiles, yeast, moulds,Pseudomonas, coliforms, LAB, and hydrogen sulfide-producing bacteria	No inoculation	0, 2, 6 and 8 d, 4°C Slices wrapped with CH or CH-5% LAE films, then packed in polyethylene films	CH films reduced 0.47–2.96 log population of fillets, depending on time and microbial group studied. Incorporation of LAE (5%) increased antimicrobial activity to 1.78–5.81 log reduction, and maintaining the initially low microbial fillets load for 8 d	[50]

Sliced turkey deli meat	Chitosan (CH, 2–5% w/w) films and coatings added with 2% solution of either acetic, lactic or levulinic acids	Lauric arginate (LAE, 50–200 mL/mL) and nisin (NIS, 25 mg/mL) alone or in combination	L. innocua (6-7 log CFU/cm2)	Even spread over the meat surface (cm2) using sterile spreaders	48 h, 37°C. Films were placed on top of inoculated turkey; coatings applied by spreading. The product was vacuum packed and stored at 10°C for 24 h prior to microbiological analysis	High CH levels reduced 4.6 log CFU/cm2. NIS addition (486 IU/cm2) reducedListeria by 2 and 2.4 log CFU/cm2 for 2% and 5% CH, respectively. Combination of CH, LAE and NIS had similar reductions as only CH with LAE, suggesting no additive or synergistic effect by NIS. Despite no statistical difference ($p<0.05$), coatings showed more microbial reduction than films	[47]

Table 3: Use of antimicrobial films and coatings in fish and seafood

Product	Coating material	Antimicrobial compound	Target microorganism	Inoculation technique	Conditions	Results	Ref.
Atlantic cod (Gadus morhua) and herring (Clupea harengus)	Chitosan coatings	Chitosan (CH) with different molecular weights and viscosities (14, 57 or 360 mPa s)	Psychrotrophic microorganisms (PT) and total plate count (TPC)	No inoculation	12 d, 4°C Samples immersed in CH solution (5°C) 30 s and after 2 min a second immersion for 30 s. Then they were dried at 40°C for 2 h in a forced air oven and stored	Herring fillets treated with 57 and 360 cP CH showed lower PT population than 14 cP CH-treated fillets after 6 d. CH treatments reduced to 103 and 102 TPC of herring and cod samples, respectively, after 12 d	[51]
Smoked salmon	Whey protein isolate (WPI) coatings	Lactoperoxidase system (LPO) (0–0.5%, w/v)	L. monocytogenes (V7 serotype 4b, LCDC 81-861 serotype 4b, Scott A serotype 4b, 101M and 108M) (102–104 CFU/g).	Spotted directly onto the salmon (I + C) or on top of applied coating (C + I) and spread with a hockey stick	4°C and 10°C, 35 d. I + C samples were inoculated and then dried for 0.5 h and coated. C + I samples dried for 1 h and then inoculated	Samples coated by LPO-WPI showed <1.0 log CFU/g of L. monocytogenesat 4°C for 35 d (both treatments). L. monocytogenes was completely inhibited in C + I samples stored during 35 d at 10°C	[54]

Cold-smoked sardine (Sardina pilchardus)	Gelatin (G) films	Oregano extract (OE) (Origanum vulgare 1.5%, v/v) or rosemary (RM) (Rosmarinus officinalis, 20% w/v) or Chitosan (CH) (1.5% w/v), high pressure (300 MPa/20°C/15 min) (HP).	Total viable count (TVC), H2S-reducing organisms, luminescent bacteria, and Enterobacteriaceae	No inoculation	5°C, 20 d Fish slices were placed between two layers of edible films and were stored in clean bags	Fish coated with OE-G and RM-G films reduced TVC by 1.99 and 1.54 log CFU/g respectively, on d 16. H2S-reducing bacteria followed a similar pattern. OE and RM had no effect, but CH reduced to ≤103 CFU/g in all cases. Pressurized samples produced undetectable levels of all microorganisms for 20 d, except uncoated sample whose TVC was 105 CFU/g at d 20	[58]
Cod (Gadus morhua)	Gelatin (G) or in combination with chitosan (CH) films	Clove essential oil (CO)	Total bacterial count (TVC), H2S-producers organisms, luminescent organisms, Pseudomonas, Enterobacteriaceae (EB), and lactic acid bacteria (LAB)	No inoculation	2°C, 11 d Fillets were covered with the G-CH film containing CO, and vacuum-packed in plastic bags	TVC count was 6.1 log CFU/g. Luminescent bacteria reached 6 log CFU/g after 3 d, but later were undetected. H2S producers were completely inhibited from d 3 onwards. LAB and EB increased during storage despite storage temperature	[48]

Cold smoked salmon (CSS) slices and fillets	Alginate (AL), -carrageenan, pectin, gelatin, or starch coatings	Sodium lactate (SL, 0–2.4% w/v) and sodium diacetate (SD, 0–0.25% w/v), OptiForm (OF, 2.5% w/v)	Mixture of L. monocytogenes strains: PSU1, PSU9, F5069, ATCC 19115, and Scott A. 103 CFU/g	Surface-inoculated	4°C, 30 d Samples coated with AL incorporating SL/SD or OF, dried 20 min and stored at 4°C in vacuum sealed bags	Al coatings with 2.4% SL/0.25% SD and OF reduced L. monocytogenesby 3.2 and 4 log CFU/g (slices) and 2.4 and 3 log CFU/g (fillets), respectively, relative to control sample	[62]
Bream (Megalobrama amblycephala)	Alginate (AL) coatings	Vitamin C (VC, 5% w/v) and tea polyphenols (TP, 0.3% w/v)	TVC	No inoculation	21 d, 4°C Bream was dipped in AL-antimicrobial solutions (1 min), air-dried (1 min) and immersed in 2% (w/v) CaCl2(1 min) to obtain gels. Samples were packed and stored	After 4 d of storage. The TVC of VC and TP decreased by 1.6 and 1.5 log CFU/g, respectively, on day 21	[63]

Sea bass slices	Gelatin extracted from the skin of unicorn leatherjacket (Aluterus monoceros) films	Lemongrass essential oil (LEO) 25% (w/w)	Mesophilic (TVC) and psychrophilic (PS) microorganisms, enterobacteria (EB), and H2S-producing bacteria LAB	No inoculation	12 d, 4°C For each slice, films were placed on both sides. Subsequently, the samples were placed in polystyrene trays wrapped with extensible polypropylene film	TVC of unwrapped sample increased to 7.2 log CFU/g at d 4 reaching 7.9 log CFU/g at d 12. TVC of LEO-film wrapped samples was 5.6 log CFU/g at d 12. PS count for control, G and LEO films was 6.0, 5.5 and 4.0 log CFU/g, respectively. LAB increased to 7.2, 6.7 and 5.9 log CFU/g at the end of storage. LEO-film showed the lowest EB counts (2.2 log CFU/g), as compared to control	[57]
Sea bass (Dicentrarchus labrax)	Chitosan (CH) films	CH with vacuum packaging	Total mesophilic aerobic bacteria (TVC) and psychrotrophic (PS) aerobic bacteria	No inoculation	4°C until end of shelf life. Fish fillets were covered using CH films, wrapped and vacuum packaged using polyethylene bags	The acceptable limit of 6 log CFU/g and 7 log CFU/g for PS and TVC bacteria, respectively, was reached after 25 d at 4°C. Control samples reached this limit after 5 d	[52]

Indian oil sardine (Sardinella longiceps)	Chitosan (CH) (1 and 2% w/v) coatings	Chitosan (1 and 2% w/v)	Mesophilic microorganisms (TVC)	No inoculation	11 d, 1-2°C. Fillets were dipped in 1 and 2% CH at 1-2°C for 10 min, drained for 5 min and placed in trays for 24 h, then sealed using HDPE	Eating quality was maintained for 8 and 10 d for 1 and 2% CH respectively, whereas untreated samples lasted 5 d. The limit of 107 CFU/g of TVC was exceeded after 7, 9 and 11 d for untreated, 1% and 2% CH treated samples, respectively	[64]
Salmon	Barley bran protein and gelatin (BBG) films	Grapefruit seed extract (GFSE) (0.5–1.2% w/v)	E. coli O157:H7 and L. monocytogenes(106CFU/mL)	E. coliO157:H7 andL. monocytogeneswere spread individually on sample surface and allowed to rest for 30 min	4°C, 15 d Samples wrapped using the BBG film. Samples packed in polyethylene terephthalate film were used as control	After 15 d, populations ofE. coli and L. monocytogenes inoculated salmon with the BBG film containing GSE decreased by 0.53 and 0.50 log CFU/g, respectively, compared to the control	[59]
Cold-smoked salmon	Potato processing waste (PPW) films	Oregano essential oil (OO) 0.97% and 1.92% (185 and 289 mg oil/g film)	L. monocytogenesV7 (6.7–6.9 log CFU/g)	L. monocytogenes Overnight culture (100 L) was spotted at 25–30 locations on salmon fillet, spread and dried in a biological hood (30 min)	4°C, 28 d Salmon samples were wrapped with edible films and were vacuum packed	Coated samples with PPW-OO, reduced Listeriapopulation by 0.4–2.4 log CFU/g as compared to control samples, after storage period	[65]

as material which can be consumed and provides a good barrier to moisture, oxygen, and solute movements for the food [14]. Because edible polymers were considered a packaging as well as a food component, they should fulfill a number of necessities, such as high barrier and mechanical efficiencies and biochemical, physicochemical, and microbial stabilities; they should be nontoxic, nonpolluting, and of low cost [15].

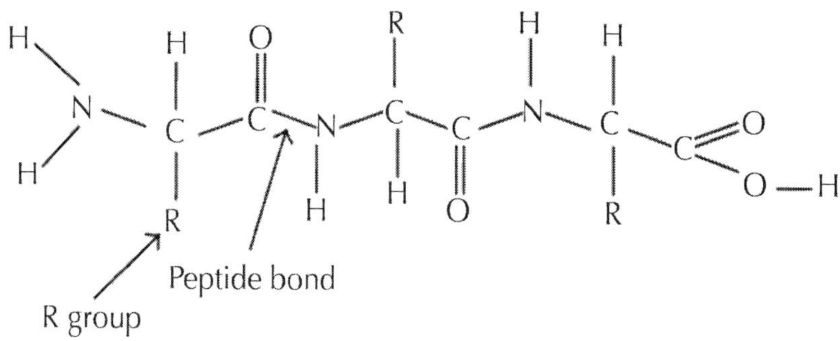

Scheme 2: Common structure of polypeptide.

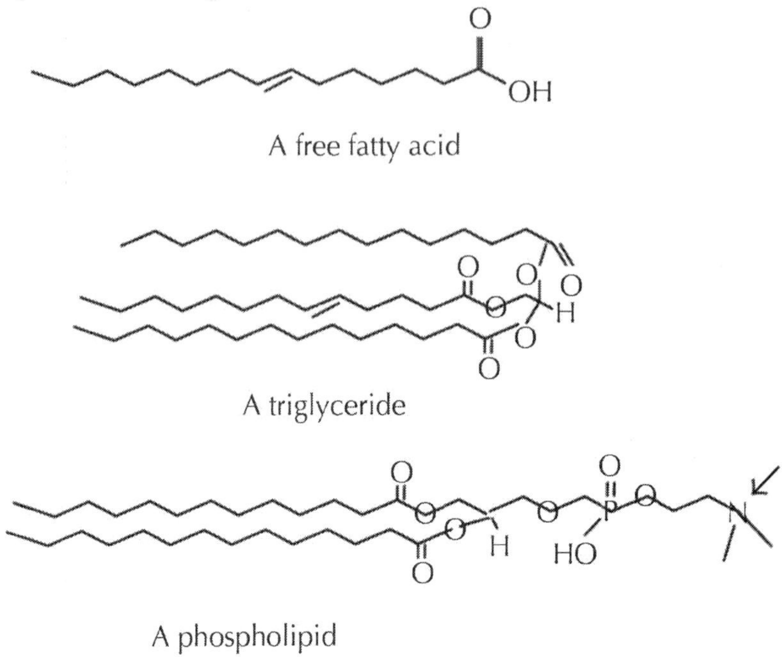

Scheme 3: Common structure of lipids.

CATEGORIES OF EDIBLE POLYMER

The edible polymers can be comprised in any of the four categories:

(1) Hydrocolloids, (2) polypeptides, (3) lipids, (4) synthetic and composite edible polymers

Hydrocolloids

Hydrocolloids are often called hydrophilic polymers, of vegetable, animal, microbial, or synthetic origin, that generally contain many hydroxyl groups (polysaccharides) and may be polyelectrolytes. Polysaccharides were commercially available for use in food and nonfood industries as stabilizers, thickening and gelling agents, crystallization inhibitors, and encapsulating agents [16]. Many hydrocolloids were polyelectrolytes, for example, starch alginate, carrageenan, carboxymethylcellulose, gum arabic, chitosan, pectin, and xanthan gum.

Starch consists of two kinds of molecules, amylose (normally 20–30%) and amylopectin (normally 70–80%), which was primarily derived from cereal grains and tubers like corn (maize), wheat, potato, tapioca, and rice. Amylose was accountable for the film forming capacity of starch [17]. High amylose starch films have been made that were flexible, oxygen impermeable, oil resistant, heat sealable, and water soluble. Polymer of high-amylose corn starch or potato starch was more stable during aging [18]. Starch-based edible polymer exhibits physical characteristics similar to plastic polymer in which they were odorless, tasteless, colorless, nontoxic, biologically absorbable, semipermeable to carbon dioxide, and resistant to passage of oxygen. Since the water activity was critical for microbial, chemical, and enzymatic activities, edible starch-based films can retard microbial growth by lowering the water activity within the package. The ability of polysaccharide-based (starch, carrageenan, and chitosan) coatings to extend the shelf life of strawberry fruit (Fragaria ananassa) was studied by Ribeiro et al., mainly for industrial applications [19]. Films, made with oxidized starch, showed improved tensile properties and fracture elongation percentage, with respect to those of the native starch [20]. Kaempferia rotunda and Curcuma xanthorrhiza are essential oils enriched in cassava starch-based edible coating which could extend patin fillet's shelf life and could be used as alternative fish preservation [21]. Alginates were derived from seaweeds and possess good film-forming properties that make them particularly useful in food applications. Alginate has a potential to form biopolymer film or coating component because of its unique colloidal properties, which include thickening, stabilizing, suspending, film forming, gel production, and emulsion stabilizing [22]. Alginate was used in various pharmaceutical preparations such as gaviscon, bisodol, and asilone. Gaviscon is a nonprescription medicine, which is taken by mouth to treat heartburn and gastroesophageal reflux disease. It was also observed that raft-forming chewable tablets prepared using an optimum amount of sodium alginate, calcium carbonate, and sodium bicarbonate could be an efficient dosage form in the treatment of gastroesophageal reflux disease [23]. 5-fluorouracil encapsulated with alginate beads was for the treatment of breast cancer [24]. Alginate based phenobarbitone tablets can be easily prepared for the treatment of chronic seizures in terms of controlled release of drug to obtain the effect over long time period [25]. Alginate was used extensively as an impression-making material in dentistry, prosthetics, life casting, and occasionally for creating positives for small-scale casting. It was also used in the food industry, for thickening soups and jellies. Edible polymer prepared from Alginates forms strong films and exhibits poor water resistance because of their hydrophilic nature [26].

Carrageenans were water-soluble polymers with a linear chain of partially sulfated galactans. These sulfated polysaccharides were extracted from the cell walls of various red seaweeds. Carrageenan film formation includes a gelation mechanism during moderate drying, leading to a three-dimensional network formed by polysaccharide-double helices and to a solid film after solvent evaporation [27]. It is used in desserts, ice cream, cream, milkshakes, salad dressings, sweetened condensed milks, and sauces, which is gel to increase viscosity. It is also used in beer, toothpaste, fruit gushers, soy milk, and so forth [28]. Recently, carrageenan films were also found to be less opaque than those made of starch [29].

Only four cellulose derivative forms are used for edible coatings or films: hydroxypropyl cellulose (HPC), hydroxypropyl methylcellulose (HPMC), carboxymethylcellulose (CMC), or methylcellulose (MC). Cellulose derivatives exhibit thermogelation; therefore when suspensions are heated they form a gel, whereas they return to original consistency when cooled [30]. The films casting from aqueous solutions of MC, HPMC, HPC, and CMC tend to have moderate strength, are resistant to oils and fats, and are flexible, transparent, flavorless, colorless, tasteless, water-soluble, and moderate barriers to oxygen. MC is more resistant to water and it is the lowest hydrophilic cellulose derivatives. However, cellulose derivative films are poor water vapor barriers because of the inherent hydrophilic nature of polysaccharides and they possess poor mechanical properties. A way to improve the moisture barrier would be the incorporation of hydrophobic compounds, such as fatty acids, into the cellulose ether matrix to develop a composite film [31]. As another example, Galus et al. [32] developed bilayers by adding a mixture of lipids (paraffin oil, paraffin wax, or a mixture of hydrogenated palm oil and triolein) onto a methylcellulose layer. Edible coatings made of CMC, MC, HPC, and HPMC have been applied to some fruits and vegetables for providing barriers to oxygen, oil, or moisture transfer.

Microcrystalline cellulose and powdered cellulose were used as inactive fillers in drug tablets and as thickeners and stabilizers in processed foods [33]. Cellulose powder was, for example, used in Kraft's Parmesan Cheese to prevent caking inside the tube. Methylcellulose (MC) and hydroxypropyl methylcellulose (HPMC) were used in coating formulations to reduce oil uptake in deep-fat frying potato strips and dough discs. MC coatings were more effective in reducing oil uptake than HPMC ones [34].

One method of enhancing the moisture barrier would be by incorporation of hydrophobic compounds such as fatty acids into the cellulose ether matrix to develop a composite film [5], for example, hydroxypropyl methylcellulose (HPMC). It was found that the moisture barrier was improved by stearic acid incorporation into the film forming solution; however, there were difficulties in preparing a homogenous composite film with both hydrophobic and hydrophilic compounds. For these reasons, homogeneous packaging film was prepared from HMPC by chemical modification allowing an increase in hydrophobicity. Cross-linking of HPMC can be considered a useful approach to prepare nonwater soluble cellulose derivatives with interesting moisture barrier properties, which could be used as biodegradable packaging materials [35].

Pectin was a group of plant-derived polysaccharides that appear to work well with low moisture foods but was poor moisture barriers. Pectin was a heterogeneous grouping of acidic structural polysaccharides, found in fruit and vegetables, and mainly prepared from citrus peel and apple pomace. The conventional application was given the jellylike consistency to jams or marmalades and jelly beans, which would otherwise be sweet juices. In medicine, pectin increases the viscosity and volume of stool so that it is used against constipation and diarrhea [36].

The agar was hydrophilic colloids that have the ability to form reversible gels simply by cooling a hot aqueous solution. Agar exhibits features that make it convenient for cutting meats. It forms strong gels characterized by melting points far above the initial gelation temperature [37]. Agar has been used comprehensively as a gelling agent in the food industry. However, despite its biodegradability and its enormous gelling power, agar has not been used widely due to poor aging. Both photodegradation and fluctuations in ambient temperature and humidity alter agar crystalline, leading to the formation of microfractures and polymer embrittlement [38]. The effect of agar on the structure and the functional properties of emulsified edible films have been recently studied by Phan et al. [39].

Chitosan was one of the auspicious natural polymers with characteristics such as biodegradability, chemical inertness, biocompatibility, high mechanical strength, good film-forming properties, and low cost [40, 41]. Additionally, chitosan was a nontoxic cellulose-like polyelectrolyte polymer hydrogel that was suitable for the production of artificial muscles, as this material undergoes a large volume change in response to changes in pH, temperature, or solvent composition [42]. Modern studies have suggested that chitin may play a role in a probable pathway in human allergic disease [43]. Chitosan was the second most abundant natural and nontoxic polymer in nature after cellulose. Some desirable properties of chitosan were that it forms films without the addition of additives and exhibits good oxygen and carbon dioxide permeability, as well as excellent mechanical properties and antimicrobial activity against bacteria, yeasts, and molds [44, 45]. Chitosan products were highly viscous, resembling natural gums. Chitosan can form translucent films to enhance the quality and extend the storage life of food products [46]. Pure chitosan films were generally cohesive and compact and the film surface has a smooth contour without pores or cracks. Chitosan was used in a wide range of applications such as wastewater treatment [47], separation membranes [48], food packaging drug delivery systems [49], and biosensors [50]. However, in sensor applications, the poor electrical conductivity of hydrogen results in a poor response time and a high operating voltage limits its applicability in devices. Hence, composites have been attempted by incorporating a rigid conducting polymer (such as PANI) into a flexible matrix (such as chitosan) to combine the good possibility of the matrix and the electrical conductivity of the conductive polymer [51]. The addition of tannic acid as a cross-linking agent of chitosan matrix has improved the mechanical properties and those related to the water affinity such as solubility, permeability, and contact angle [52].

Gums in edible-polymer preparation were used for their texturizing capabilities. In edible-forming preparations, guar gum was used as a water binder, stabilizer, and viscosity builder. Gum arabic, owing to its solubility in hot or cold water, was the least viscous one of the hydrocolloid gums. Xanthan gum was readily dispersed in water; hence high consistency was obtained rapidly in both hot and cold systems. A blend of guar gum, gum Arabic, and xanthan gum provided uniform coatings with good cling and improved adhesion in wet batters. The mesquite gum forms films with excellent water vapor (Tables 1 and 2) barrier properties when small amounts of lipids were added in their formulation [52].

Polypeptides

Polypeptide- (protein-) based edible polymer can be used for the individual packaging of small portions of food, particularly products that were not currently individually packaged for practical reasons, such as beans, nuts, and cashew nuts. Additionally, polypeptide-based edible polymer can function as carriers for antimicrobial and antioxidant agents. Additional possible application of polypeptide-based edible polymer could be their use of multilayer food packaging materials together with nonedible polymer. In this case, the protein-based edible polymer would be the internal layers in direct contact with food materials. Through functions relating to mechanical and barrier properties, protein-based edible polymer may be able to substitute synthetic polymer films.

Polypeptides were anticipated to be good oxygen barriers at low relative humidity. However, the fact that proteins were not totally hydrophobic and contain predominantly hydrophilic amino acid residues limits their moisture-barrier properties. Collagens were the major structural proteins of connective tissue, such as bone, hide, tendons cartilage, and ligaments. The main application of collagen films was as barrier membrane [53]. Films with a thickness of 0.01–0.5 mm are formed by air-drying a casted collagen preparation similar to ophthalmological shields and made of biodegradable materials. The drugs can be loaded into collagen membranes by hydrogen bonding, covalent bonding, or simple entrapment. They can be sterilized and become made flexible. This was used in ophthalmology in delivering drugs to yes and also used over wounds to make healing rapid and uniform [54]. Collagen polymer offers several advantages; it was biocompatible and nontoxic to most tissues; it has well-documented structural, physical, chemical, and immunological properties; it can be processed into a variety of forms, and it is readily isolated and purified in large quantities [55, 56]. Gelatin was unique in forming a thermoreversible substance with a melting point close to body temperature, which was mainly significant in edible and pharmaceutical applications. Fundamentally, gelatin was obtained by controlling hydrolysis from the fibrous insoluble protein, collagen, which was widely found in nature as the major constituent of skin, bones, and connective tissue. The characteristic features of gelatin were the high content of the amino acids glycine, proline, and hydroxyproline. Gelatin also has a mixture of single and double unfolded chains of a hydrophilic character. At approximately 40°C, gelatin aqueous solutions were in the sol state and form physical, thermoreversible gels on cooling. Gelatin was used to encapsulate low moisture or oil phase food constituents and pharmaceuticals [6]. Such encapsulation provides fortification against oxygen and light, as well as defining the amount of ingredient or drug dosage. In addition, gelatin films have been formed as coatings on meats to reduce oxygen, moisture, and transport of oil [57]. Addition, gelatin was competent to form clear and strong films and was used for microencapsulation and capsule coatings in food and pharmaceutical manufacturing. Microencapsulation is the process of surrounding or enveloping one substance within another substance on a very small scale, yielding capsules ranging from less than one micron to several hundred microns in size. Substances may be microencapsulated with the intention that the core material is confined within capsule walls for a specific period of time. Alternatively, core materials may be encapsulated so that the core material will be released either gradually through the capsule walls, known as controlled release or diffusion, or when external conditions trigger the capsule walls to rupture, melt, or dissolve [58]. Short chain gelatin peptides are effective in the treatment of stomach ulcers.

Zein was a prolamin protein and therefore dissolves in 70–80% ethanol [59]. Zein was a relatively hydrophobic and thermoplastic material [60]. Zein coating has also shown an ability to reduce moisture and loss of firmness and delay color alteration in fresh fruit like apples [61]. Furthermore, zein may similarly take part in the coating of conventional packaging plastics. Zein's properties make it valuable in processed foods and pharmaceuticals, in competition with insect shellac. It was now used as a crust for candy, nuts, fruit, pills, and other encapsulated foods and drugs [62]. Wheat gluten was a water insoluble protein of wheat flour which was comprised of a mixture of polypeptide molecules and considered to be globular proteins. The cohesiveness and elasticity of the gluten give reliability to wheat dough and expedites film formation. The poor resistance of wheat gluten films to water vapor (Tables 1 and 2) was due to the hydrophilic nature of the protein and the substantial amount of hydrophilic plasticizer added to impart acceptable film flexibility. The properties of wheat gluten films can be upgraded by using a cross-linking agent such as glutaraldehyde or heat curing. Wheat gluten films can be abundantly biodegraded after 36 days in aerobic fermentation and within 50 days in farmland soil without liberating toxic products. It can be used in a wide selection of applications including milling, bakery products, meats, pasta, and breads [63]. Soy protein from soybeans has been extensively used as a food ingredient in nearly every food product available to the consumer, since it contains high nutrition and excellent functional properties. Since soy proteins were abundant, inexpensive, biodegradable, and nutritional, they show the potential to be developed as edible and biodegradable films [64, 65]. Milk proteins can be classified into two types: casein and whey proteins. The acid casein can be converted to functional soluble caseinates by neutralization through the addition of alkali. Edible protein films based on various caseinates can be obtained by solubilization in water followed by casting and drying. Laminated films that comprised casein did protect dried fruit and vegetables from moisture absorption and oxidation [12]. Caseinate-lipid emulsion coatings were fruitful in plummeting moisture loss from skinned carrots and zucchini [66].

Mung beans were of interest as a potential component of biopolymeric films because of their high protein content [67]. Bourtoom [68] synthesized and investigated the films from mung bean protein. It was found that the mechanical properties (tensile strength and elongation at break) of mung bean protein films had superior mechanical properties and water vapor barrier properties (Tables 1 and 2). These were better than other protein sources such as casein, soy protein isolate, wheat gluten, peanut proteins, and water-soluble fish protein films. However, the mung bean protein films still showed substantially lower mechanical and water vapor barrier properties (Tables 1 and 2), compared with some synthetic polymers (high density polyethylene, polyvinyl chloride, cellulose acetate, and polyester).

soluble starch, hydroxypropyl starch and gelatin, corn zein and corn starch [72], gelatin and fatty acid, soy protein isolates and gelatin [73], and soy protein isolate and polylactic acid [74].

Polyvinyl acetate was a nontoxic commercially important polymer prepared through emulsion polymerization which has been investigated as a coating film containing fungicides for protection of diverse foods and as a coating for pharmaceutical products. Biopolymers such as polysaccharides and proteins or even biocompatible synthetic polymers such as polymethyl methacrylate (PMMA) could be better alternatives for combinations with antimicrobial polymers aiming at antimicrobial chemotherapy in vivo. The use of antimicrobial macromolecular agents, however, was not restricted to in vivo applications; they may also be used ex vivo for water disinfection, food packing and preservation, and many antifouling applications [75]. For more than 50 years, techniques such as compression spray and dip coating and encapsulation have been used in the pharmaceutical industry to incorporate bioactive agents with polymers. Such polymers have largely included cellulose derivatives, polyethylene glycol (PEG), and poly (N-vinylpyrrolidone). Peppas and coworkers have initiated the use of pH-responsive complexation hydrogels of poly (methacrylic acid) grafted with PEG, referred to as P (MAA-g-EG), for oral protein delivery. Micelle-forming polymers, such as block copolymers of poly (ethylene oxide) and poly (propylene oxide), or Pluronics, have been thoroughly studied in drug delivery. Polyplexes formed by cooperative electrostatic interactions between polyethyleneimine (PEI) and DNA were widely studied for gene delivery [76].

An anionic copolymer based on methacrylic acid and methyl methacrylates was used for the coating of tablets and pills which resistance to gastric juice and improve the keeping property protection against moisture, light, and air coatings stable under tropical conditions [77]. Polymer spheres with a sugar coating on the outside and plastic coating on the inside have been made by European scientists. This gives them dual functionality to target and deliver drugs. Helmut Schlaad, from the Max Planck Institute of Colloids and Interfaces, Potsdam, and colleagues from Germany and Switzerland made the spheres by dissolving glycosylated polybutadiene-poly (ethylene oxide) block copolymers in water. When dissolved, the copolymers spontaneously formed hollow colloids called vesicles with a glucose coating on the outside and a poly (ethylene oxide) coating on the inside. The polymer vesicles could be used as living cell mimics or drug delivery vessels. They could be used to target drugs and biomolecules to injured or cancerous tissues [78].

SPECIAL APPLICATIONS OF EDIBLE POLYMERS

Heightened consumer demand for enhanced keeping quality and freshness of foods has given rise to the concept of active packaging—a type of packaging that alters conditions surrounding the food to maintain product quality and freshness, improve sensory properties, or enhance product safety and shelf life. To meet the growing demand of recyclable or natural packaging materials and consumer demands for safer and better quality foods, new and novel food-grade packaging materials or technologies have been and continue to be developed. Examples of these packaging materials include biobased polymers, bioplastic or edible polymer packaging products made from raw materials originating from agricultural or marine sources [79]. The application of nanocomposites promises to expand the use of edible and biodegradable polymers [80]. It will help to reduce the packaging waste associated with processed foods and will support the preservation of fresh foods, extending their shelf life. Polymer composites were mixtures of polymers with inorganic or organic additives having certain geometries (fibers, flakes, spheres, and particulates). The use of nanoscale fillers was leading to the development of polymer nanocomposites and represents a radical alternative to these conventional polymer composites [81]. Pharmaceutical and food products were often coated with edible shellac, which must be of a sufficiently reduced viscosity. The application of polyvinyl acetate (PVA) was assessed as a potential protective edible coating in round-type tomato fruit on the green stage of maturity. The edible polymeric coating did not significantly affect brightness, weight loss, and RR; however, environmental conditions at which the fruits were maintained had significant effects. Despite the statistically nonsignificant effects, results showed that the PVA coating was associated with slight protection of the fruit compared to uncoated fruits [82]. Gelatin capsules [83] were small shells made from gelatin that were used to enclose various medications and supplements. Capsules can make it easier to take certain drugs and supplements, especially those in a powder or liquid form. The gelatin was almost odorless and tasteless, and it typically does not cause digestive problems for most people. Gelatin also usually was not interacting with other drugs. Gelatins from alternatives to mammalian species were gaining prominence, especially gelatins from marine fish species. Because of their good film-forming abilities, fish gelatins may be a good alternative to synthetic plastics for making films preserve foodstuffs [84].

Edible shellac coatings preserve fruits and vegetables during their long trek from the farm to your local produce stand. To improve these coatings, Agricultural Research Service (ARS) scientists developed a new coating made of food-grade polyvinyl acetate that is cheaper to use and more effective than shellac at preventing postharvest fruit decay without discoloring the fruit. ARS has applied for a patent for this breakthrough [85]. It was cheaper to use and more effective than shellac at preventing postharvest fruit decay without discoloring the fruit. The coating can be applied to fruits and vegetables by dipping, spraying, or brushing on. Polyvinyl acetate (PVA) was a synthetic polymer that was used as an ingredient in chewing gum [86]. In addition to the polyvinyl acetate, the coating for fresh produce can include plasticizers, surfactants that aid coverage, gloss-enhancing additives, and other ingredients for specific uses, such as coating candy and baked goods. The new coating has several advantages over shellac, which slows fruit respiration and keeps the fruit firm [85]. Shellac tends to whiten or "blush" when it was wide open to moisture. This often occurs when apples were moved from cold storage to a humid environment. Another problem was that citrus fruits and some apple varieties develop "off" flavors when coated with shellac. And the fact that shellac coatings are primarily composed of insect exudates has made them objectionable to some consumers.

Several examples of snacks of orally soluble edible polymer, including their compositions and manufacturing techniques, were disclosed [87]. The snacks may include one or more layers of film that was orally soluble and fragments quickly upon placement in a human mouth without leaving substantial residue that can be felt by the human tongue or which needs to be swallowed or ejected from the mouth. An edible snack unit of tasty chocolate-flavored orally soluble polymer is comprised of a base medium formed in the shape of an ingestible snack, said base medium being formed at least in part from an orally soluble polymer, and said snack being orally soluble such that it disintegrates relatively slowly when placed on the roof of a human mouth and licked [88]. These film-forming edible polymers can include pullulan, hydroxypropyl methylcellulose, hydroxyethyl cellulose, hydroxypropyl cellulose, polyvinyl pyrrolidone, carboxymethyl cellulose, polyvinyl alcohol, sodium alginate, polyethylene glycol, xanthan gum, tragacanth gum, guar gum, acacia gum, gum arabic, polyacrylic acid, methyl methacrylate copolymer, carboxyvinyl polymer, amylose, high amylose starch, hydroxypropylated high amylose starch, dextrin, pectin, chitin, chitosan, levan, elsinan, collagen, gelatin, zein, gluten, soy protein isolate, whey protein isolate, casein, polysaccharides, natural gums, polypeptides, polyacrylates, starch, gum karaya, gelatin, mixtures thereof, and others. Chewing gum consists of a gum base, sugar, corn syrup, softeners, and flavorings. The gum base was the insoluble part left in the mouth while chewing. The gum base was made of resins from trees, latexes, or the milky juices from plants, and man-made polymers. Thomas Adams desired to use the chicle as a rubber substitute, but instead, rubber was used to make chewing gum—unvulcanized styrene-butadiene synthetic rubber (SBR), others utilize polyvinyl acetate polymer (PVA) [89]. Modern chewing gum was made with a synthetic rubber, polyisobutylene, which not only makes it elastic, but also contributes to the obstinate, sticky quality. US 3761286 [90] reveals a chewing gum composition comprising chewing gum and a water insoluble hydrophilic polymer of a hydroxyalkyl acrylate or methacrylate having controllably, releasable adsorbed there in a flavor, wherein the hydrophilic polymer can be poly-(2-hydroxy ethyl methacrylate). Gebreselassie et al. [91] invented swellable gum compositions which increase in volume upon chewing. The gum compositions include a super absorbent hydrophilic polymer in combination with a gum base or as a component of a gum base.

Researchers around the globe were trying to find ways of improving therapeutic efficacy of drugs by modifying the formulation technique, polymeric systems, and so forth. The use of novel edible polymers offers benefits. Care should be taken to properly select polymers while designing a delivery system. The eventual goal was to introduce cost effective, biocompatible, multifunctional, and less toxic polymers so that the delivery systems pass through the various phases of clinical trials and benefit the society. It was believed that the advances in polymer sciences will transfigure the design, development, and performance of polymer based drug delivery systems. PCL and PLA were chemically for bioactive polymeric compounds. Nano composites perception represents a motivating route for creating new and innovative materials, also in area of edible polymers. Materials with a large variety of properties have been realized, and even more due to be realized. Micro- and Nano encapsulation of active compounds with edible polymer coatings may help to control their release under specific conditions, thus protecting them from moisture, heat, or other extreme conditions and enhancing their stability and viability. Coating foods with Nano laminates involves either dipping them into a series of solutions containing substances that would be adsorbed to a food's surface or spraying substances onto the food surface. These Nano laminate coatings could be elaborated entirely from food-grade ingredients and could include various functional agents such as antimicrobials, antibrowning agents, antioxidants, enzymes, flavorings, and colorant. In fact, the layer-by- layer electrode position technique [101] could be used to coat highly hydrophilic food systems such as fresh-cut fruits and vegetables including further vitamins and antimicrobial agents. The Nano composite materials obtained by mixing natural, edible polymers, and sheets of crystalline solid layered offer a great variety of property profile. They were even able to compete, both in price and in performance, with synthetic polymeric materials. Consumer demands were driving research and development for alternatives to petroleum-based packaging materials including those with recyclable or edible properties, as well as those materials made from renewable/ sustainable agricultural products. Edible films, gels, or coatings were considered biopolymers with numerous desirable properties and may be made from a variety of materials, including polysaccharides, lipids, and proteins, alone or in combination with other components. Edible biopolymers also have been developed from other sources and applied to foods, including fungal exopolysaccharides (pullan) or fermentation byproducts (polylactic acid).

CONCLUSIONS

The practice of edible polymers on many food products continues to develop. The potential paybacks of edible polymers as carriers of antimicrobial agents, flavors, antioxidants, coloring agents, vitamins, probiotics, and nutraceutical excuse continued research in this field of active packaging. A new trend in the storage of foods and in the food industry, nowadays, was the use of recyclable, renewable agricultural products for the production of packages, edible films, and coatings. Edible polymer technology plays an energetic role in drug discovery and was improving outcomes for patients today and addressing unmet therapeutic needs in the future. Vaccine drug delivery systems were now being proven to be patient friendly as they avoid the need to administer booster doses and provide a long term therapy in small doses. Edible polymer vaccines on the other hand open an attractive avenue for the oral delivery of vaccines. The development of new technologies to improve the delivery properties of edible polymer was a major issue for future research. At that moment, most studies on food applications have been conducted at a laboratory scale. However, further research should be focused on a commercial scale with the purpose of providing more accurate information that can be used to commercialize fresh cut products coated with edible polymers. Food industries were looking for edible polymers that could be used on a broad spectrum of foods and add value to their products while increasing their shelf life. When the active ingredients (antimicrobials, antioxidants, and nutrients) were added to edible polymer, mechanical, sensory, and even functional properties can be dramatically affected. Scientists have researched new production methods to make coatings edible, safe, and fully functional. According to a broad range of research findings, edible polymer may be applied to prolong the shelf life of food products, control material exchange, and improve the products sensory properties, nutritive value, and attractiveness. Edible biopolymers were competitive as both commodity polymers and specialty applications where a specific functionality was valued. In a scenario where petroleum-based polymers were replaced by biopolymers, edible biopolymers from food crops would primarily be used in food applications, whereas other biopolymers could well cover the demand for commodity polymers. A continued development of functionality and processing was needed for a commercial breakthrough.

REFERENCES

1. S. Kokoszka and A. Lenart, "Edible coatings—formation, characteristics and use—a review," Polish Journal of Food and Nutrition Sciences, vol. 57, no. 4, pp. 399–404, 2007.

2. G. G. Buonocore, M. A. del Nobile, A. Panizza, S. Bove, G. Battaglia, and L. Nicolais, "Modeling the lysozyme release kinetics from antimicrobial films intended for food packaging applications," Journal of Food Science, vol. 68, no. 4, pp. 1365–1370, 2003.

3. R. C. Eberhart, S. Su, K. T. Nguyen et al., "Bioresorbable polymeric stents: current status and future promise," Journal of Biomaterials Science, Polymer Edition, vol. 14, no. 4, pp. 299–312, 2003.

4. D. Bikiaris, "Nanomedicine in cancer treatment: drug targeting and the safety of the used materials for drug nanoencapsulation," Biochemical Pharmacology, vol. 1, article E122, 2012.

5. A. Dhanapal, P. Sasikala, L. Rajamani, V. Kavitha, G. Yazhini, and M. S. Banu, "Edible films from polysaccharides," Food Science and Quality Management, vol. 3, pp. 9–18, 2012.

6. T. Bourtoom, "Edible films and coatings: characteristics and properties," International Food Research Journal, vol. 15, no. 3, pp. 237–248, 2008.

7. World Economic Situation and Prospects 2012, Sales no. E.12. Ii.C.2, United Nations Publication, 2012.

8. E. Salleh, I. I. Muhamad, and N. Khairuddin, "Structural characterization and physical properties of antimicrobial (AM) starch-based films," World Academy of Science, Engineering and Technology, vol. 3, pp. 7–25, 2009.

9. Z. Akbari, T. Ghomashchi, and S. Moghadam, "Improvement in food packaging industry with biobased nanocomposites," International Journal of Food Engineering, vol. 3, no. 4, article 3, 2007.

10. S. Saremnezhad, M. H. Azizi, M. Barzegar, S. Abbasi, and E. Ahmadi, "Properties of a new edible film made of faba bean protein isolate," Journal of Agricultural Science and Technology, vol. 13, no. 2, pp. 181–192, 2011.

11. O. Skurtys, C. Acevedo, F. Pedreschi, J. Enronoe, F. Osorio, and J. M. Aguilera, Food Hydrocolloid Edible Films and Coatings, Food Science and Technology, Nova Publisher, 2010.

12. M. E. Embuscado and K. C. Huber, Edible Films and Coatings for Food Applications, Springer, London, UK, 2009.

13. S. A. Jang, G. O. Lim, and K. B. Song, "Preparation and mechanical properties of edible rapeseed protein films," Journal of Food Science, vol. 76, no. 2, pp. C218–C223, 2011.
14. P. Mokrejs, F. Langmaier, D. Janacova, M. Mladek, K. Kolomaznik, and V. Vasek, "Thermal study and solubility tests of films based on amaranth flour starch-protein hydrolysate," Journal of Thermal Analysis and Calorimetry, vol. 98, no. 1, pp. 299–307, 2009.
15. P. Bergo, P. J. A. Sobral, and J. M. Prison, "Effect of glycerol on physical properties of cassava starch films," Journal of Food Processing and Preservation, vol. 34, no. 2, pp. 401–410, 2010
16. A. M. Stephen, G. O. Phillips, and P. A. Williams, Food Polysaccharides and Their Applications, CRC Press, Taylor & Francis Group, 2nd edition, 2006.
17. C. A. Romero-Bastida, L. A. Bello-Pérez, M. A. García, M. N. Martino, J. Solorza-Feria, and N. E. Zaritzky, "Physicochemical and microstructural characterization of films prepared by thermal and cold gelatinization from non-conventional sources of starches," Carbohydrate Polymers, vol. 60, no. 2, pp. 235–244, 2005.
18. K. Krogars, J. Heinämäki, M. Karjalainen, J. Rantanen, P. Luukkonen, and J. Yliruusi, "Development and characterization of aqueous amylose-rich maize starch dispersion for film formation," European Journal of Pharmaceutics and Biopharmaceutics, vol. 56, no. 2, pp. 215–221, 2003.
19. C. Ribeiro, A. A. Vicente, J. A. Teixeira, and C. Miranda, "Optimization of edible coating composition to retard strawberry fruit senescence," Postharvest Biology and Technology, vol. 44, no. 1, pp. 63–70, 2007.
20. E. Argüello-García, J. Solorza-Feria, J. R. Rendón-Villalobos, F. Rodríguez-González, A. Jiménez-Pérez, and E. Flores-Huicochea, "Properties of edible films based on oxidized starch and zein," International Journal of Polymer Science, vol. 2014, Article ID 292404, 9 pages, 2014.
21. R. Utami, Kawiji, E. Nurhartadi, A. Y. T. Putra, and I. Setiawan, "The effect of cassava starch-based edible coating enriched with Kaempferia rotunda and Curcuma xanthorrhiza essential oil on refrigerated patin fillets quality," International Food Research Journal, vol. 21, no. 1, pp. 413–419, 2014.

42. S. J. Kim, S. R. Shin, S. M. Lee, I. Y. Kim, and S. I. Kim, "Electromechanical properties of hydrogels based on chitosan and poly(hydroxyethyl methacrylate) in NaCl solution," Smart Materials and Structures, vol. 13, no. 5, pp. 1036–1039, 2004.
43. C. G. Lee, "Chitin, chitinases and chitinase-like proteins in allergic inflammation and tissue remodeling,"Yonsei Medical Journal, vol. 50, no. 1, pp. 22–30, 2009. · ·
44. J. Vartiainen, R. Motion, H. Kulonen, M. Rättö, E. Skyttä, and R. Ahvenainen, "Chitosan-coated paper: effects of nisin and different acids on the antimicrobial activity," Journal of Applied Polymer Science, vol. 94, no. 3, pp. 986–993, 2004. · ·
45. S.Y. Park, K. S. Marsh, and J. W. Rhim, "Characteristics of different molecular weight chitosan films affected by the type of organic solvents," Journal of Food Science, vol. 67, no. 1, pp. 194–197, 2002.
46. C. Ribeiro, A. A. Vicente, J. A. Teixeira, and C. Miranda, "Optimization of edible coating composition to retard strawberry fruit senescence," Postharvest Biology and Technology, vol. 44, no. 1, pp. 63–70, 2007.
47. G. Crini, "Non-conventional low-cost adsorbents for dye removal: a review," Bioresource Technology, vol. 97, no. 9, pp. 1061–1085, 2006.
48. W. Won, X. Feng, and D. Lawless, "Pervaporation with chitosan membranes: separation of dimethyl carbonate/methanol/water mixtures," Journal of Membrane Science, vol. 209, no. 2, pp. 493–508, 2002.
49. J. Nunthanid, M. Laungtana-Anan, P. Sriamornsak et al., "Characterization of chitosan acetate as a binder for sustained release tablets," Journal of Controlled Release, vol. 99, no. 1, pp. 15–26, 2004.
50. Y. Tsai, S. Chen, and H. Liaw, "Immobilization of lactate dehydrogenase within multiwalled carbon nanotube-chitosan nanocomposite for application to lactate biosensors," Sensors and Actuators B: Chemical, vol. 125, no. 2, pp. 474–481, 2007.
51. X. H. Xu, G. L. Ren, J. Cheng, Q. Liu, D. G. Li, and Q. Chen, "Self-assembly of polyaniline-grafted chitosan/glucose oxidase nanolayered films for electrochemical biosensor applications,"

Journal of Materials Science, vol. 41, no. 15, pp. 4974–4977, 2006.

52. R. Díaz-Sobac, H. García, C. I. Beristain, and E. J. Vernon-Carter, "Morphology and water vapor permeability of emulsion films based on mesquite gum," Journal of Food Processing and Preservation, vol. 26, no. 2, pp. 129–141, 2002.

53. R. Dimitriou, G. I. Mataliotakis, G. M. Calori, and P. V. Giannoudis, "The role of barrier membranes for guided bone regeneration and restoration of large bone defects: current experimental and clinical evidence," BMC Medicine, vol. 10, article 81, 2012.

54. R. Khan and M. H. Khan, "Use of collagen as a biomaterial: an update," Journal of Indian Society of Periodontology, vol. 17, no. 4, pp. 539–542, 2013.

55. K. Dybka and P. Walczak, "Collagen hydrolysates as a new diet supplement," Food Chemistry and Biotechnology, vol. 73, pp. 83–92, 2009.

56. T. Wittaya, "Protein-based edible films: characteristics and improvement of properties," in Structure and Function of Food Engineering, chapter 3, InTech, 2012.

57. M. I. Khan, M. N. Adrees, M. R. Tariq, and M. Sohaib, "Application of edible coating for improving meat quality: a review," Pakistan Journal of Food Sciences, vol. 23, no. 2, pp. 71–79, 2013.

58. J. Sris, A. Seethadevi, K. Suria Prabha, P. Muthuprasanna, and P. Pavitra, "Microencapsulation: a review," International Journal of Pharma and Bio Sciences, vol. 3, no. 1, pp. 509–521, 2012.

59. T. Wittaya, "Protein-based edible films: characteristics and improvement of properties," In Structure and Function of Food Engineering, chapter 3, InTech. ·

60. R. Shukla and M. Cheryan, "Zein: the industrial protein from corn," Industrial Crops and Products, vol. 13, no. 3, pp. 171–192, 2001.

61. J. Bai, V. Alleyne, R. D. Hagenmaier, J. P. Mattheis, and E. A. Baldwin, "Formulation of zein coatings for apples (Malus domestica Borkh)," Postharvest Biology and Technology, vol. 28, no. 2, pp. 259–268, 2003.

62. H. Dhanya and S. Divia, "Development of Zein-Pectin nanoparticle as drug carrier," International Journal of Drug Delivery, vol. 4, pp. 147–152, 2012.

63. M. M. Moore, T. J. Schober, P. Dockery, and E. K. Arendt, "Textural comparisons of gluten-free and wheat-based doughs, batters, and breads," Cereal Chemistry, vol. 81, no. 5, pp. 567–575, 2004.
64. S. Y. Cho and C. Rhee, "Mechanical properties and water vapor permeability of edible films made from fractionated soy proteins with ultrafiltration," Lebensmittel-Wissenschaft und Technologie, vol. 37, no. 8, pp. 833–839, 2004.
65. J. Zhang, P. Mungara, and J. Jane, "Mechanical and thermal properties of extruded soy protein sheets,"Polymer, vol. 42, no. 6, pp. 2569–2578, 2001.
66. Bisson, "Preparation of a Casein-Based Puffed Product," United States Patent. Patent Number: 4,744,99, May 1988.
67. S. Keereekasetsuk and T. Bourtoom, "Influence of plasticizers on the properties of edible film from Mung bean protein," in Proceedings of the 14th World Congress of Food Science and Technology, Shanghai, China, October 2008.
68. T. Bourtoom, "Factors affecting the properties of edible film prepared from mung bean proteins,"International Food Research Journal, vol. 15, no. 2, pp. 167–180, 2008.
69. F. F. Shih, K. W. Daigle, and E. T. Champagne, "Effect of rice wax on water vapour permeability and sorption properties of edible pullulan films," Food Chemistry, vol. 127, no. 1, pp. 118–121, 2011.
70. S. Berg, M. Bretz, E. M. Hubbermann, and K. Schwarz, "Influence of different pectins on powder characteristics of microencapsulated anthocyanins and their impact on drug retention of shellac coated granulate," Journal of Food Engineering, vol. 108, no. 1, pp. 158–165, 2012.
71. S. L. Kamper and O. N. Fennema, "Water vapor permeability of an edible, fatty acid, bilayer film,"Journal of Food Science, vol. 49, pp. 1482–1485, 1984.
72. S. Y. Ryu, J. W. Rhim, H. J. Roh, and S. S. Kim, "Preparation and physical properties of zein-coated high-amylose corn starch film," LWT—Food Science and Technology, vol. 35, no. 8, pp. 680–686, 2002.
73. N. Cao, Y. Fua, and Y. He, "Preparation and physical properties of soy protein isolate and gelatin composite films," Lwt—Food Science and Technology, vol. 35, pp. 680–686, 2002.

74. J. W. Rhim, A. Gennadios, A. Handa, C. L. Weller, and M. A. Hanna, "Solubility, tensile, and color properties of modified soy protein isolate films," Journal of Agricultural and Food Chemistry, vol. 48, no. 10, pp. 4937–4941, 2000.

75. A. M. Carmona-Ribeiro and L. D. D. M. Carrasco, "Cationic antimicrobial polymers and their assemblies," International Journal of Molecular Sciences, vol. 14, no. 5, pp. 9906–9946, 2013.

76. S. Yamano, J. Dai, S. Hanatani et al., "Long-term efficient gene delivery using polyethylenimine with modified Tat peptide," Biomaterials, vol. 35, no. 5, pp. 1705–1715, 2014.

77. H. U. Petereit, C. Meier, and E. Roth, United States Patent. Patent no. : Us 7,160,558 B2, January 2007.

78. H. Schlaad, L. You, R. Sigel et al., "Glycopolymer vesicles with an asymmetric membrane," Chemical Communications, no. 12, pp. 1478–1480, 2009.

79. D. S. Cha and M. S. Chinnan, "Biopolymer-based antimicrobial packaging: a review," Critical Reviews in Food Science and Nutrition, vol. 44, no. 4, pp. 223–237, 2004.

80. L. Lagarón Cabedo, D. Cava, J. Feijoo, R. Gavara, and E. Gimenez, "Improving packaged food quality and safety, part 2: nanocomposites," Food Additives and Contaminants, vol. 22, no. 10, pp. 994–998, 2005.

81. E. W. Gacitua, A. A. Ballerini, and J. Zhang, "Polymer nanocomposites: synthetic and natural fillers a review," Maderas: Ciencia y Tecnólogía, vol. 7, no. 3, pp. 159–178, 2005.

82. G. Y. Cortez-Mazatán, L. A. Valdez-Aguilar, R. H. Lira-Saldivar, and R. D. Peralta-Rodríguez, "Polyvinyl acetate as an edible coating for fruits. Effect on selected physiological and quality characteristics of tomato," Revista Chapingo Serie Horticultura, vol. 17, no. 1, pp. 15–22, 2011.

83. A. Lodha, A. Patel, J. Chaudhuri, P. Jadia, T. Joshi, and J. Dalal, "Formulation and evaluation of transparent ibuprofen soft gelatin capsule," Journal of Pharmacy and Bioallied Sciences, vol. 4, pp. 95–97, 2012.

84. M. C. Gómez-Guillén, M. Pérez-Mateos, J. Gómez-Estaca, E. López-Caballero, B. Giménez, and P. Montero, "Fish gelatin: a

renewable material for developing active biodegradable films," Trends in Food Science and Technology, vol. 20, no. 1, pp. 3–16, 2009.

85. K. Grohmann and R. D. Hagenmaier, "Edible Food Coatings Containing Polyvinyl Acetate," United States Patent. Patent Number: 6,162,475, December 2000.
86. P. R. Gunjal, T. S. Kalmegh, M. Gadhave, and S. A. Jadhav, "Review on medicated chewing gum as a novel drug delivery system," International Journal of Universal Pharmacy and Life Sciences, vol. 2, no. 4, pp. 19–36, 2012.
87. T. M. Pearce and T. V. Pearce, "Snacks of Orally Soluble Edible Films," United States Patent. Pub. no. : Us 2003/0224090 A1, December 2003.
88. T. Pearce, "Delivery Units of Thick Orally Soluble Polymer," United States Patent Application Publication. Pub. no. : Us 2004/0247746 A1, December 2004.
89. E. John, Vanity, Vitality, and Virility, pp. 189–197, Oxford University Press, New York, NY, USA, 2004.
90. T. H. Shepherd and F. E. Gould, "Chewing Gum Composition Containing Flavored Homopolymers," United States Patent Office. Patent no. 3,761,286, September 1973.
91. P. Gebreselassie, N. Boghani, and G. Visscher, "Edible Compositions Containing Swellable Polymers," United States Patent Application Publication. Pub. no. : Us 2006/0153949 A1, July 2006.
92. D. R. Lu, C. M. Xiao, and S. J. Xu, "Starch-based completely biodegradable polymer materials," Express Polymer Letters, vol. 3, no. 6, pp. 366–375, 2009.
93. V. G. Kadajji and G. V. Betageri, "Water soluble polymers for pharmaceutical applications," Polymers, vol. 3, pp. 1972–2009, 2011.
94. Y. Luo and Q. Wang, "Zein-based micro- and nano-particles for drug and nutrient delivery: a review,"Journal of Applied Polymer Science, 2014.
95. J. Necas, L. Bartosikova, P. Brauner, and J. Kolar, "Hyaluronic acid (hyaluronan): a review," Veterinarni Medicina, vol. 53, no. 8, pp. 397–411, 2008.

96. V. T. Chamle, P. S. Misal, S. G. Shep, and R. A. Hajare, "A potential natural gum as a polymer used in NDDS: recent investigations," International Journal of Current Trends in Pharmaceutical Research, vol. 1, no. 1, pp. 67–73, 2013.
97. I. J. Ogaji, E. I. Nep, and J. D. Audu-Peter, "Advances in natural polymers as pharmaceutical excipients,"Pharmaceutica Analytica Acta, vol. 3, no. 1, article 146, 2012.
98. G. Vilar, J. Tulla-Puche, and F. Albericio, "Polymers and drug delivery systems," Current Drug Delivery, vol. 4, pp. 367–394, 2012.
99. Y. J. Kim, S. Chun, J. Whitacre, and C. J. Bettinger, "Self-deployable current sources fabricated from edible materials," Journal of Materials Chemistry B, vol. 1, no. 31, pp. 3781–3788, 2013.
100. B. Mattix, T. R. Olsen, T. Moore et al., "Accelerated iron oxide nanoparticle degradation mediated by polyester encapsulation within cellular spheroids," Advanced Functional Materials, vol. 24, no. 6, pp. 800–807, 2014.
101. P. Jochenweiss and D. Julianmcclements, "Functional materials in food nanotechnology," Journal of Food Science, vol. 71, no. 9, pp. 10–116, 2006.

Chapter 4

Food Analysis: Present, Future, and Foodomics

Alejandro Cifuentes

Laboratory of Foodomics, Institute of Food Science Research (CIAL), CSIC, Nicolas Cabrera 9, Campus de Cantoblanco, 28049 Madrid, Spain

ABSTRACT

This paper presents a revision on the instrumental analytical techniques and methods used in food analysis together with their main applications in food science research. The present paper includes a brief historical perspective on food analysis, together with a deep revision on the current state of the art of modern analytical instruments, methodologies, and applications in food analysis with a special emphasis on the works published on this topic in the last three years (2009–2011). The article also discusses the present and future challenges in food analysis, the application of "omics" in food analysis (including epigenomics, genomics, transcriptomics, proteomics, and metabolomics), and provides an overview on the new discipline of Foodomics.

FOOD ANALYSIS: A BRIEF HISTORICAL PERSPECTIVE

Analysis of foods is continuously requesting the development of more robust, efficient, sensitive, and cost-effective analytical methodologies to guarantee the safety, quality, and traceability of foods in compliance with legislation and consumers' demands. The old methods used at the beginning of the 20th century based on the so-called "wet chemistry" have evolved into the current powerful instrumental techniques used in food laboratories. This improvement has led to significant enhancements in analytical accuracy, precision, detection limits, and sample throughput, thereby expanding the practical range of food applications. As mentioned by McGorrin [1] "the growth and infrastructure of the modern global food distribution system heavily relies on food analysis (beyond simple characterization) as a tool for new product development, quality control, regulatory enforcement, and problem-solving." Besides, currently, there is also a huge interest in the health-related properties of foods as a result of an increasing public concern on how to improve health through the so-called functional foods, functional ingredients, and nutraceuticals. Thus, there is no doubt on the importance and current need of analytical techniques developments able to face all these demands.

Traditionally, analytical techniques have been classified according to their working principle. For example, they can be spectroscopic (e.g., mass spectrometry (MS); nuclear magnetic resonance (NMR); infrared (IR); atomic spectroscopy (AS)), biological (polymerase chain reaction (PCR); immunological techniques; biosensors), electrochemical (including also biosensors here), for separation (e.g., high-performance liquid chromatography (HPLC); gas chromatography (GC); capillary electrophoresis (CE); supercritical fluid chromatography

(SFC)), for sample preparation (e.g., solid phase extraction (SPE); supercritical fluid extraction (SFE); headspace (HS); flow injection analysis (FIA); purge and trap (PAT); microwave-assisted extraction (MAE); automatic thermal desorption (ATD)), hyphenated (e.g., putting together separation and spectroscopic techniques), and so forth. Every technique provides specific information on the sample or components under study based on a specific physical-chemical interaction, and all have their own advantages and drawbacks when applied to food analysis as will be discussed below. A description of the huge number of analytical techniques commonly used in food analysis is out of the scope of this work. An additional idea on the complexity of the number of techniques currently involved in food analysis can be obtained developing a little more one of the above subdisciplines. Considering the case of immunological techniques, they include the following ones: enzyme immunoassay (EIA), enzyme-linked immunosorbent assay (ELISA), immunodotting, radioimmunoassay (RIA), solid-phase RIA, liquid-phase RIA, immunoradiometric assay (IRMA), fluorescence: fluorescence immunoassay, enzyme-linked fluorescent immunoassay, fluorescence polarization immunoassay (FPIA), time-resolved fluorescence immunoassay (TRFI), and chemiluminescence immunoassay (CIA), up to 27 techniques can be associated to immunological techniques.

The present paper will focus on recent developments and applications of modern instrumental analytical techniques to detect compounds of significance to food science and technology, with a special emphasis on the literature published from January 2009 to January 2012. Namely, the present work will focus on modern analytical instrumentation, the development of new methods and their application in food science and technology including the new field of Foodomics [2], as well as recent works on quality control and safety, nutritional value, processing effects, storage, bioactivity, and so forth. The analysis of a large variety of food-related molecules with different chemical properties, including amino acids, peptides, proteins, phenolic compounds, carbohydrates, DNA fragments, vitamins, toxins, pesticides, additives, and chiral compounds, is also described below. Besides, a global picture on the evolution of the principal analytical techniques employed in food analysis will be provided through a comparison between the number of works published on food analysis in the first ten years of this 21st century (2001–2011) with the number of works published in the last ten years of the past 20th century (1990–2000) on the same topic.

FOOD ANALYSIS: CURRENT STATE OF THE ART, METHODOLOGIES, AND APPLICATIONS

A large number of works have directly focused on the analytical technique used in food analysis, while others have focused on the type of food, compound, or process investigated. Regarding specific analytical techniques applied to solve different problems in food analysis, one of the more active areas is the development of sample preparation techniques, in good agreement with the complex nature of foods. Sample preparation is one of the key steps for the development of any new analytical methodology; as a result, research on new sample preparation procedures is one of the most active areas in analytical chemistry. Advances in sample preparation aim to minimize laboratory solvent use and hazardous waste production, save employee labor and time, and reduce the cost per sample, while improving the efficiency of the analyte isolation. This includes the development and application of molecularly imprinted polymers in food sample analysis [3], the use of monoliths in sample preparation and analysis of milk [4], the development of porous monolith microextraction techniques for determination of veterinary residues in food matrices by [5], the use of the so-called QuEChERS (quick, easy, cheap, effective, rugged, and safe) methodology for determining pesticide residues in food matrices [6], the application of immunoaffinity column clean-up techniques in food analysis [7], the development of solid-phase microextraction (SPME) techniques for quality characterization of food products [8], the application of ultrasound-assisted extraction to the determination of contaminants in food and soil samples [9], and the use of liquid phase microextraction in food analysis [10]. Besides, some papers have focused on the description of sample preparation strategies used for the analysis of aflatoxins in food and feed [11], antibacterial residues in foodstuffs [12], or the determination of pesticides in foods [13]. At present, new green sample preparation methods are being studied; among them, supercritical fluid extraction (SFE) and subcritical water extraction (SWE, also called accelerated solvent extraction) are among the more promising processes in food science, not only in food analysis [14] but also for obtaining new functional food ingredients.

These extraction techniques based on pressurized fluids provide higher selectivities, shorter extraction times, and are environmentally friendly.

Following with the description based on specific analytical techniques, separation techniques continue as one of the more active areas in food analysis. Separations based on liquid chromatography (LC) are the most numerous as can be deduced from the many works published on the use of different formats of LC [15, 16], including hydrophilic interaction liquid chromatography (HILIC) [17, 18], nano-LC [10], or high-speed counter-current chromatography [19], just to name a few. Other techniques as gas chromatography (GC) still have an important role in food analysis, for instance, for the analysis of volatile fractions or fatty acids in foods [20]. Electrodriven separation techniques such as capillary electrophoresis (CE) or microchip capillary electrophoresis have found important applications in food analysis as can be deduced from the many review works devoted to this topic [21, 22], including the detection of genetically modified organisms [23, 24], nucleosides and nucleotides in foods [25], analysis of contaminants in emerging food safety issues and food traceability [26], and food-borne pathogens [27].

It is also interesting to realize how mass spectrometry (MS) has evolved in the last years in food analysis. In the 1990s, MS was mostly used as GC detector to confirm the identification of analytes. During the last decade, MS has mainly been used for direct identification and quantification of food compounds typically coupled to other separation techniques like LC and, in a less extent, CE. Single quadrupole MS has been restricted to screening purposes since these instruments do not meet the more recent criteria set by food regulators as FDA or EFSA, especially those regarding the requested number of identification points. As a result, tandem MS has become a general tool for the identification and quantification of analytes (mainly contaminants) in food analysis. The enhanced selectivity afforded by tandem MS detection may also contribute to the simplification of the extraction procedure if attention is paid to ion suppression phenomena. At this point, the use of triple quadrupole, ion trap, and more recently time of flight MS analyzers coupled to uni- or bidimensional separation techniques have been widely reported in the scientific literature in food analysis [28, 29]. Other MS applications include analysis of pesticides and their metabolites in food and water matrices [30, 31], analysis of food proteins and peptides [32], MS-based analytical methodologies to characterize genetically modified crops [33], MALDI-TOF MS analysis of plant proanthocyanidins [34], or multistage mass spectrometry in quality, safety, and origin of foods [35]. It is expected that new developments on ionization techniques prior to MS analysis can make even broader its application in food analysis [36].

Spectroscopic techniques are based on the principle that molecules and atoms can interact with electromagnetic radiation. Structural, physical-chemical, and/or qualitative-quantitative information of the compounds under study can thus be obtained. This information is provided by the wavelength or frequency detected in the emitted or absorbed energy spectrum. Spectroscopic techniques have found in food analysis a large use due to that they are fast, give direct measurement of the food constituents, do not use toxic reactants and solvents, can be used in process line, are not destructive and noninvasive, and some of them can detect several compounds simultaneously [37]. Nowadays, spectroscopic techniques based on infrared region are one of the most numerous in the food analysis. Thus, infrared spectroscopy is frequently used for quality control of food including analysis of honey [38] or muscle food [39]. Spectroscopic techniques based on the near infrared (NIR) spectrum have also been used to identify transgenic foods [40] or for measuring bioactive compounds in foods [41]. The use of the midinfrared region has been applied to study the secondary structure of food proteins [42] or to study intact food systems exploring their molecular structure-quality relationships [43], while raman spectroscopy has also found a number of applications in agricultural products and food analysis [44]. Another spectroscopic technique such as nuclear magnetic resonance (NMR) has been used for the rapid analysis of oil and fact content in agrifood products [45], or in the mode 31P NMR to solve different problems in food analysis [46]. Other topics currently under development are the application of chemiluminescence in food analysis [47] or the use of chemiluminescence as detection in LC [48] or CE [49]. Other spectroscopic techniques like fluorescence [50] or modern electrothermal atomic absorption spectrometry have also found interesting applications in food analysis [51].

Hyphenated techniques have found a huge number of possibilities in food analysis. This is the case of liquid chromatography online coupled to mass spectrometry (LC-MS) or tandem MS (LC-MS/MS) which have been extensively applied to ensure food safety [52, 53], particularly to analyze antimicrobials residues in food of animal origin [54], antibiotics in food samples [55], clenbuterol residues [56], food allergens [57], and so forth. Other hyphenated techniques such as gas chromatography-mass spectrometry (GC-MS) or capillary electrophoresis-mass spectrometry (CE-MS) have also found interesting applications to analyze essential oils [58] or food contaminants [59].

Biological techniques employ living organisms or some of their products such as enzymes, antibodies, and DNAs, to identify and analyze foods. Although biological-type techniques such as the classical enzymatic or microbiological analysis will not be specifically covered in this paper, their use is still a must in many microbiological and food laboratories. The main developments in this area have been brought about by the important advances in the following.

- DNA-based techniques and molecular methods have allowed the fast and sensitive detection of Salmonella in foods [60], the detection of fraudulent seafood species substitution [61], or the microbial composition of different foods [62, 63]. The detection of multiple genetically engineered crops [64] is still an important topic in which the effect of food processing on plant DNA degradation and PCR-based analysis of transgenic foods has also been studied [65]. These DNA-based methods also allow the authenticity determination of meat and meat products [66], the identification of animal species in food products [67], as well as the authentication of maize [68].
- Biosensors are analytical devices composed of a biological recognition element (e.g., enzyme, antibody, and microbe) coupled to a chemical or physical transducer that converts the chemical signal into an electrical response. Biosensors provide a continuous or semicontinuous electronic signal proportional to an analyte or group of analytes. Biosensors have been developed for determining major and minor food components, preservatives, food colors and sweeteners, toxins, pesticides, antibiotics, and hormones [69]. They have also found use in tracking microbial contamination [70], to follow food safety [71], processing, and to certify food quality and control including the development of the so-called electronic nose [72] or electronic tongue [73] that have a large impact on flavor analysis. The advantages of these methods are the rapid response time, the high degree of specificity and sensitivity, and the possibility of being used for inline processes monitoring food manufacturing.

- Other developments include the use of peptide nucleic acid (PNA)-based technologies for food analysis and food authentication [74] or the development of new immunoassay methodologies for analyzing veterinary drug residues in foods and food products [75] or to characterize plant food allergens [76].

On the other hand, many other works—instead of focusing on a given analytical technique for food analysis as above—have studied specific food constituents for whose analysis different analytical techniques and methods have been developed. One of the main topics is the case of contaminants in foods whose analysis is a must for ensuring human exposure to noxious residues though diet does not exceed acceptable levels for health. As mentioned above, the everyday more exigent demands on food safety are bringing about a tremendous development in analytical instruments and methodologies to analyze foodborne pathogens [77] and contaminants [78–80], or the effect of food processing on pesticide residues in fruits and vegetables [81]. Good examples of the tremendous work devoted to these issues are the analysis of mycotoxins in foods [82,83], the analysis of other toxin-like compounds as, for instance, sterigmatocystin, a toxic metabolite closely related to aflatoxins, and produced by the fungi Aspergillus nidulans and A. versicolor that has been found in different foods [84]. Another important issue is the analysis of antibiotics in foods [85], including different legal aspects related to their determination and subsequent validation of the analytical methodologies [86]. Robust analytical methods are continuously under development in order to improve the figures of merit (analysis speed, resolution, and sensitivity), allowing the fast and sensitive analysis of other food contaminants as well as possible dangerous compounds generated during food processing as acrylamide [87,88], melatonin [89], N-nitrosamines [90], dioxins and dioxin-like PCBs [91], polycyclic aromatic hydrocarbons [92], Sudan dyes [93], food taints and off-flavours [94], thiols in wine [95], and sulphites in food and beverages [96].

Currently there is an important amount of work devoted to the study of food not only considered as source of energy, but also as a natural source of valuable ingredients than can provide additional health benefits. Following this trend new terms as functional foods, functional ingredients, or nutraceuticals are now in use in many laboratories that investigate links between the composition of food and its health benefits. This has been the case of isoflavones that has been largely analyzed in food and biological fluids based on their claimed health benefits supposedly due to their antioxidant activity [97, 98], a point that is now under controversy. Similar approach has been applied for the advanced analysis of food anthocyanins, isoflavones, and flavanols [99], flavonoids [100], phenolic compounds [101, 102], carotenoids [103, 104], and other phytochemicals in foods [105]. Many times these studies focus on the biological properties of other compounds, as fatty acids [106] or glucosinolates [107] that can have other sources different than plants. The analysis of nutraceuticals is a key issue for which advanced analytical techniques will play a key role [108] as well as for global research in nutrition and dietetics [109]. In this regard, other important topics are the determination of bioactive compounds in cranberry [110] or the determination of amino acids in grape-derived products [111]. Some compounds have also attracted much attention due to their beneficial rheological properties, this is the case of the analysis of the gelling carrageenans [112], or the determination of the degree of N acetylation of the polymers chitin and chitosan [113].

Other more specific applications in food analysis have also seen a great development as a result of the combination of several analytical advances that have been put together. A good example is the study of the geographical origin of foods via the analyses of stable isotope ratio of light elements [114] or the meta-analysis of the effects of pasteurization on milk vitamins [115]. This is also the case of the analysis of the volatile fraction of foods, which is known to have a crucial effect on food quality and acceptance. The study of the volatile fraction of food or beverage requires analytical methods and technologies able not only to evaluate its composition exhaustively but also to monitor variations of its profile and to detect trace components characterizing the food being investigated. The strategies of analysis have changed significantly over the last 15–20 years because of the introduction of new approaches, in particular: (i) solventless sample preparation techniques; (ii) fast gas chromatography and related techniques; (iii) new analytical techniques, such as comprehensive gas chromatography (GC); (iv) new operative strategies based on approaches developed for other fields and applied to food analysis; (v) data elaboration strategies producing a higher level of information [116]. Chiral analysis has also seen an important growing in food analysis, since chiral methods can be used to study and characterize foods and beverages through the enantiomeric separation of different food compounds such as amino acids, pesticides, and polyphenols [117]. Another example is the investigation on food texture in which physical characteristics perceived by the senses are investigated. Research in this area has evolved tremendously in the last decade based on multidisciplinary approaches that encompass chemistry, physics, physiology, and psychology, to study fracture of food, the sounds it makes during biting and chewing, its microstructure, muscle movements during mastication, swallowing, and acceptability, and so forth [118, 119].

Several books have also been printed in this period focusing on molecular techniques for detection and characterization of foodborne pathogens [120, 121], molecular biological and immunological techniques for food chemists [122], mass spectrometry in food safety [123], food analysis instruments [124], dairy food analysis [125], grains identity preservation and traceability [126], food contaminants [127], antibiotic residues in food [128], and fortified foods with vitamins [129].

Also, several book chapters have focused on the use of FTIR for rapid authentication and detection of adulteration of food [130], time-domain NMR applied to food products [131], or biosensors for functional food safety and analysis [132]. Readers interested in this topic can also take resort of several special issues published by different journals on food analysis [133], advanced separation methods in food analysis [134], allergens in foods [135], natural bioactive compounds and nutrigenomics [136], food and beverage analysis [137] and advanced food analysis [138].

Table 1 provides information on the number of works published in the period 2001–2011 found through a search in the database Food Science and Technology Abstracts (FSTA) using as key terms the names of the analytical technique indicated in each case

Table 1: Number of works published in the period 2001–2011 found through a search in the database Food Science and Technology Abstracts (FSTA) using as key terms the names of the analytical technique indicated in each case

Technique	Number of ref.
Sample preparation	

Microwave-assisted extraction	384
Headspace	2571
Solid phase extraction	3866
Supercritical fluid extraction	680
Purge and trap	151
Flow injection analysis	393
Pressurized liquid extraction	436
Microextraction	2201
Biological	
Biosensors	750
PCR	7085
Microbiological analysis	416
Recombinant DNA	220
Immunological techniques	3008
Othersa	118
Separation	
Liquid chromatography	8927
Gas chromatography	4798
SDS/PAGE	3227
Capillary electrophoresis	1155
Supercritical fluid chromatography	51
LC LC, LC-LC	27
GC GC, GC-GC	210
LC-GC	38
Spectroscopic	
Mass spectrometry	5030
Fluorescence	4807
NMR	3735
Infrared	2369
X-ray	2119
Ultraviolet	1429
Atomic spectroscopy	1046
Electron spectroscopy	1026
Light scattering	891
Circular dichroism	468

Othersb	932
Rheological	
Creep	205
Oscillatory shear	203
Rheometry	195
Viscometry	163
Stress relaxation	145
Normal stress	32
Thermal	
DSC	551
Thermogravimetry	17
The mochemical	16
Differential thermal analysis	9
Radiochemical	
Radioimmunoassay	96
Isotopic	140
Radiochemical	31
Radiometric	18
Radioisotope	8
Radiotracer	5
Radiolabelling	2
Electrochemical	
Biosensors	750
Voltammetry	532
Potentiometry	234
Amperometry	245
Polarography	47
Conductometry	43
Coulometry	32

[a]The group "others" includes radioimmunoassay and enzymatic analysis.

[b]The group "others" includes raman (402), electron spin resonance (366), dielectric spectroscopy (57), refractometry (54), polarimetry (38), chemiluminescence (15), and photoacoustic (0).

There are some important conclusions that can be extracted from the results of Table 1 when they are compared to the results from a similar search published by our group at the end of the 20th century summarizing the works published on food analysis in the period 1990–2000 [139]. The most important trend is the huge increase in biological and sample preparation techniques as compared with the previous period and the important decrease in the use of radiochemical and thermal techniques, probably due to the specific information that those techniques provide and the need for high-throughput techniques widely based on new and advanced technologies able to provide more information of better quality. Thus, it is not strange that techniques such as thermal and radiochemical have decreased by half (compared to the previous period) and others such as spectroscopic, biological, and sample preparation techniques have increased two, three, and four times, respectively. Other well-established techniques such as separation techniques continue to be used in a high extend, but nowadays they are not the most widely used (as in the period 1990–2000), since spectroscopic techniques have gained importance and at present are the most extensively used in food analysis.

PRESENT CHALLENGES IN FOOD ANALYSIS

Currently there are a good number of challenges to be solved in food analysis. The variety of toxic residues in food is continuously increasing as a consequence of industrial development, new agricultural practices, environmental pollution, and climate change. A critical issue will be how to detect untargeted compounds and determine their identity in foods, for which the development of advanced analytical techniques is expected to play a crucial role [140]. This increasing number of food contaminants is bringing about the development of everyday more powerful, sensitive, and fast analytical methodologies able to detect emerging contaminants in food-like industrial organic pollutants, nanomaterials, pharmaceutical residues antibiotics, and coccidiostats or emerging groups of marine biotoxins [141]. In spite of these important developments, still hundreds of foodborne infection cases occur around the world, and up to one third of the population in industrialized nations suffers from foodborne illness each year. Microbiologists have developed over the last decades reliable culture-based techniques for pathogens detection in foods.

These methods are considered to be the "gold-standard"; however, they remain cumbersome and time consuming. The introduction of genetic-based technologies makes feasible developing sensitive and specific screening tests for the detection of microbial pathogens. Microarray-based technologies represent an advance in nucleic acid testing methods whose main features include miniaturization, ability to parallelize sample processing, and ease of automation [142, 143]. Despite the advent of these rapid detection methods based on molecular techniques (or immunoassays), it is suggested that reduction and/or elimination of cultural enrichment will be essential in the quest for truly real-time detection methods. As such, there is an important role for the so-called preanalytical sample processing that in this case would include bacterial concentration and purification from the sample matrix as a step preceding detection [144]. In this regard, one analytical challenge that still remains in food safety is to present reliable results with respect to official guidelines, as fast as possible without impairing method properties such as recovery, accuracy, sensitivity, selectivity, and specificity [78].

It is also now under discussion the way in which the toxicity of food chemical contaminants is addressed. It seems it should be optimized considering food cooking, processing and eating habits, parameters that usually are neglected [145]. In addition, there is also a need of new and alternative (in vivo) toxicity tests as well as new detection tools for testing the effects of food chemical contaminants, moving away from the mouse bioassay for food toxin analysis [146]. To achieve this goal, the use of noncarcinogenic functional cell models, of alternative animal models like zebrafish embryos, and a toxicogenomic approach, seem to be the most promising strategies for the future toxicity assessment of food chemical contaminants [147].

There is also a good amount of discussion on how to ensure food safety around the globe and the many roles of risk analysis including microbial risk assessment [148, 149]. In this regard, benefit-risk assessment in food and nutrition weighs the beneficial and adverse effects that a food (component) may have, in order to facilitate more informed management decisions regarding public health issues. Although benefit-risk assessment is now in its infancy, and its exact scope is yet to be defined, benefit-risk assessment can be a valuable approach to provide the best possible science-based answer to complicated questions with a large potential impact on public health [150]. To achieve these goals it will be necessary to develop new and more powerful tools for the performance assessment and improvement of food safety following a global approach [151]. For instance, a good topic to evaluate through benefit-risk assessment will be the use of nanomaterials in food science. Nanotechnology and nanomaterials have remarkable potential to enhance the food supply through novel applications, including nutrient and bioactive absorption and delivery systems; microbial, allergen, and contaminant detection and control; food packaging properties and performance; improved colors and flavors. Based on these multiple applications, exposure to nanomaterials in the human food chain may occur not only through intentional uses in food manufacturing, but also via uses in agricultural production and carry over from use in other industries. New analytical methods are, therefore, needed to fully detect and characterize nanomaterials incorporated into foods and in other media. Moreover, there is also a need for additional toxicology studies on different types of nanomaterials to understand how they can affect food safety [152]. Besides, nanomaterials are also expected to play a crucial role in the miniaturization of analytical systems [153] including newly emerging technologies able to offer platforms with greater automation and multiplexing capabilities [154] or in the development of new nanomaterial-based biosensors for food analysis applications [155]. These multiplexed bioanalytical techniques are expected to provide control agencies and food industries with new possibilities for improved, more efficient monitoring of food, and environmental contaminants. In this regard, developments in planar-array and suspension-array technologies have demonstrated their potential in detecting pathogens, food allergens and adulterants, toxins, antibiotics, and environmental contaminants [156]. In this context, microfluidics technology has also shown interesting applications for food analysis although more effort has to be put on the development of multipurpose microfluidic platforms that integrate multiple unit operations for real food sample analysis [157]. Miniaturized systems and their applications are expected to keep growing in food analysis.

More suitable analytical techniques are still required for the detection of allergens in foods because minute amounts of the allergen can have critical consequences in sensitized persons. Although immunological methods are currently preferred, the determination of allergenic proteins by LC-MS has advanced in recent years, and it is now frequently used for the identification and quantitation of food allergens [57]. In spite of these advances, confirmatory alternatives are still needed to be able to face other additional problems originated, for example, by food matrix interferences or food processing, which may not influence allergenicity but do impair allergen detection.

Regarding comprehensive and multidimensional chromatography methods (e.g., LC × LC; GC × GC), an important point to consider is that nowadays these multidimensional techniques require dedicated laboratories, equipment, and highly trained personnel, what can explain the relatively low number of applications of these techniques [159–161]. In order to improve the simplicity and speed of analysis provided by these techniques, keeping the cost of analysis as low as possible, important instrumental developments have to be achieved. They include the improvement of the connection of the two systems to overcome the relatively costly operation conditions in GC × GC or the loss in sensitivity in LC × LC. In the coming years, new solutions should appear in order to facilitate these couplings as well as to further increase the orthogonality of the systems and, consequently, their separation power and applications in food analysis [159].

Nowadays, food laboratories are being pushed to exchange their classical procedures for modern analytical techniques that allow them to give an adequate answer to global demands on food safety, quality, and traceability. Besides, the Montreal Protocol has had a major impact also on food laboratories that have been pushed to develop more environmental-friendly methods. An important challenge is the implementation of Green Analytical Chemistry [162] also in food analysis laboratories [163] including, for instance, the greening of sample preparation techniques (with the use of new green solvents, miniaturization, or employment of solventless techniques) and the combination with new (and cleaner) separation techniques.

FOOD ANALYSIS IN THE POSTGENOMIC ERA: OMICS APPROACHES AND FOODOMICS

The sequencing of nearly the whole human genome at the beginning of 21st century is considered the beginning of the so-called postgenomic era. The advances observed in this period have made feasible to count on analytical instruments and methodological developments that were unthinkable few decades ago. As a result, researchers in food science and nutrition are being pushed to move from classical methodologies to more advanced strategies usually borrowing methods well established in medical, pharmacological, and/or biotechnology research. In this context, we coined and defined Foodomics as a discipline that studies the food and nutrition domains through the application of advanced omics technologies to improve consumer's well-being, health, and knowledge [2, 21, 164]. The main idea behind the use of this new term has been not only to use it as a flag of the new times for food analysis, but also to highlight that the investigation into traditional and new problems in food analysis in the postgenomic era can find exciting opportunities and new answers through the use of epigenomics, genomics, transcriptomics, proteomics, and metabolomics tools. A description on the basis of these omics approaches is given next.

Epigenomics studies the mechanisms of gene expression that can be maintained across cell divisions, and thus the life of the organism, without changing the DNA sequence. The epigenetic mechanisms are related to changes induced (e.g., by toxins or bioactive food ingredients) in gene expression via altered DNA methylation patterns, altered histone modifications, or noncoding RNAs, including small RNAs. The global analysis of gene expression as approached by transcriptomics offers impressive opportunities in Foodomics (e.g., for the identification of the effect of bioactive food constituents on homeostatic regulation and how this regulation is potentially altered in the development of certain chronic diseases). Two conceptually different analytical approaches have emerged to allow quantitative and comprehensive analysis of changes in mRNA expression levels of hundreds or thousands of genes. One approach is based on microarray technology, and the other group of techniques is based on DNA sequencing. Next, typically real-time-PCR is applied to confirm the up- or downregulation of a selected number of genes. In proteomics, the huge dynamic concentration range of proteins in biological samples causes many detection difficulties due to that many proteins are below the sensitivity threshold of the most advanced instruments. For this reason, fractionation and subsequent concentration of the proteome are often needed. Besides, the use and development of high-resolving separation techniques as well as highly accurate mass spectrometers are nowadays essential to solve the proteome complexity. Two fundamental analytical strategies can be employed: the bottom-up and the top-down approach. Both methodologies differ on the separation requirements and the type of MS instrumentation. New proteomic approaches based on array technology are also being employed. Metabolome can be defined as the full set of endogenous or exogenous low molecular weight metabolic entities of approximately <1000 Da (metabolites), and the small pathway motifs that are present in a biological system (cell, tissue, organ, organism, or species). Unlike nucleic acid or protein-based omics techniques, which intend to determine a single chemical class of compounds, metabolomics has to deal with very different compounds of very diverse chemical and physical properties. Moreover, the relative concentration of metabolites in the biological fluids can vary from millimolar level (or higher) to picomolar, making it easy to exceed the linear range of the analytical techniques employed. No single analytical methodology or platform is applicable to detect, quantify, and identify all metabolites in a biological sample. Two analytical platforms are currently used for metabolomic analyses: MS- and NMR-based systems. These techniques either stand alone, or combined with separation techniques (typically, LC-NMR, GC-MS, LC-MS, and CE-MS), can produce complementary analytical information to attain more extensive metabolome coverage. There are three basic approaches that can be used in metabolomics: target analysis, metabolic profiling, and metabolic fingerprinting. Target analysis aims the quantitative measurement of selected analytes, such as specific biomarkers or reaction products. Metabolic profiling is a nontarget strategy that focuses on the study of a group of related metabolites or a specific metabolic pathway. Meanwhile, metabolic fingerprinting does not aim to identify all metabolites, but to compare patterns of metabolites that change in response to the cellular environment.

Due to the huge amount of data usually obtained from omics studies, it has been necessary to develop strategies to convert the complex raw data obtained into useful information. Thus, bioinformatics has become also a crucial tool in Foodomics. Over the last years, the use of biological knowledge accumulated in public databases by means of bioinformatics allows to systematically analyse large data lists in an attempt to assemble a summary of the most significant biological aspects. Also, statistical tools are usually applied, for example, for exploratory data analysis to determine correlations among samples (which can be caused by either a biological difference or a methodological bias), for discriminating the complete data list and reduce it with the most relevant ones for biomarkers discovery, and so forth

The application of advanced omics technologies as proposed by Foodomics has already found application in different topics in food science and nutrition. Some examples are given below. Proteomics and metabolomics represent powerful analytical strategies to acquire more detailed and complete information on food composition even beyond the traditional food component analysis. Following this idea, proteomic methods have already been applied to several issues related to food safety [165], as well as the optimization of processing and quality control of food of animal origin [166], to investigate the milk proteome [167, 168], the rice proteome [169], or to study allergy and intolerance to wheat products [170]. Transcriptomic, proteomic, nd metabolomic approaches are also valuable tools to distinguish between similar food products and to detect food frauds (adulteration, origin, authenticity, etc.), food-borne pathogens, toxic species, food allergens, and so forth. For instance, in the context of food safety and transcriptomics, several DNA microarray chips have already been developed for the detection of food-borne pathogens, toxigenic microorganisms, GMO analysis, and so forth. Proteomic and metabolic changes also occur during crops growing conditions, food processing/preparation (fermentation, baking, boiling, etc.), and food conservation/storage (freezing, smoking, drying, etc.). These tools have already been demonstrated to be very useful for getting a deeper understanding of molecular details of foods and food related matrices [171–173] including the analysis of GM foods. In this later case, the use of omics approaches able to provide useful fingerprints of GM foods (e.g., for GM detection, composition monitoring, traceability, study of unintended modifications, and labelling issues) has already been recommended by the European Food Safety Authority [33, 174–176]. In this context, metabolomics (via GC-MS, LC-MS, CE-MS, or NMR) has potential to add significant value to crop and food science, raw material quality and safety, food storage, shelf life, and postharvest processing [177]. There are also other topics in which Foodomics can play a crucial role including the comprehensive assessment of food safety, quality and traceability ideally as a whole [164]; the investigation on the global role and functions of gut microbiome, a topic that is expected to open an impressive field of research [178, 179]; the stress adaptation responses of food-borne pathogens to ensure food hygiene, processing and preservation [180], or the molecular basis of biological processes with agronomic interest and economic relevance, such as the interaction between crops and its pathogens, as well as physicochemical changes that take place during fruit ripening [181].

Currently, a main area of research in food science is the connection between food and health. Today, food is considered not only a source of energy but also an affordable way to prevent future diseases. The number of opportunities (e.g., new methodologies, new generated knowledge, new products, etc.) derived from this trend are impressive, and it includes, for example, the possibility to account for food products tailored to promote the health and well-being of groups of population identified on the basis of their individual genomes. Interaction of modern food science and nutrition with disciplines such as pharmacology, medicine, or biotechnology provides impressive new challenges and opportunities. As a result, advanced analytical methodologies, "omics" approaches, and bioinformatics—frequently together with in vitro, in vivo, and/or clinical assays—are applied to investigate topics in food science and nutrition that were considered unapproachable few years ago demonstrating the huge possibilities of Foodomics in this important area of research [182]. However, food scientists and nutritionists have to face a large number of challenges to adequately answer the new questions emerging from this new field of research. One of the main challenges is to improve our limited understanding of the roles of nutritional compounds at molecular level (i.e., their interaction with genes and their subsequent effect on proteins and metabolites) for the rational design of strategies to manipulate cell functions through diet, which is expected to have an extraordinary impact on our health. The problem to solve is huge, and it includes the study of the individual variations in gene sequences, particularly in single nucleotide polymorphisms (SNPs), and their expected different answer to nutrients. Moreover, nutrients can be considered as signalling molecules that are recognized by specific cellular-sensing mechanisms [183]. However, unlike pharmaceuticals, the simultaneous presence of a variety of nutrients, with diverse chemical structures and concentrations and having numerous targets with different affinities and specificities, increases enormously the complexity of the problem [184]. Therefore, it is necessary to look at hundreds of test compounds simultaneously and observe the diverse temporal and spatial responses. Besides, several of the health benefits assigned to many dietary constituents are still under controversy as can be deduced from the large number of applications rejected by the European Food Safety Authority about health claims of new foods and ingredients [185]. More sounded scientific evidences are needed to demonstrate or not the claimed beneficial effects of these new foods and constituents. In this sense, the advent of new postgenomic strategies as Foodomics seems to be essential to understand how the bioactive compounds from diet interact at molecular and cellular level, as well as to provide better scientific evidences on their health benefits. The combination of the information from the three expression levels (gen, protein, and metabolite) can be crucial to adequately understand and scientifically sustain the health benefits from food ingredients. A representation of an ideal Foodomics strategy to investigate the effect of food ingredient(s) on a given system (cell, tissue, organ, or organism) is shown in Figure 1. Following this Foodomics strategy, results on the effect of food ingredient(s) at genomic/transcriptomic/proteomic and/or metabolomic level are obtained, making possible new investigations on food bioactivity and its effect on human health at molecular level [186].

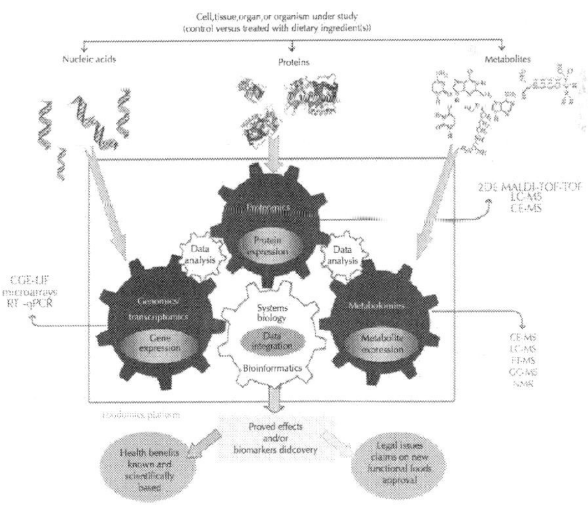

Figure 1: Scheme of an ideal Foodomics platform to investigate the health benefits from dietary constituents on a given biological system (cell, tissue,

organ, or organism), including analytical methodologies used and expected outcomes. Modified from [158] with permission from Elsevier.

However, in spite of the significant outcomes expected from global Foodomics strategies, there is still a long way to go since practically there are no papers published in literature in which results from the three expression levels (transcriptomics, proteomics, and metabolomics) are simultaneously presented and merged being the number of review papers on this topic higher than the number of research papers. To our knowledge, only two papers have presented results from the three expression levels simultaneously following a global strategy [158, 187]. Figure 2 shows the results from one of these global Foodomics studies on the chemopreventive effect of dietary polyphenols against HT29 colon cancer cells proliferation [158]. Figure 2shows the genes, proteins and metabolites that were identified (after transcriptomic, proteomic, and metabolomic analysis) to be involved in the principal biological processes altered in HT29 colon cancer cells after the treatment with rosemary polyphenols, showing the huge potential of this global strategy.

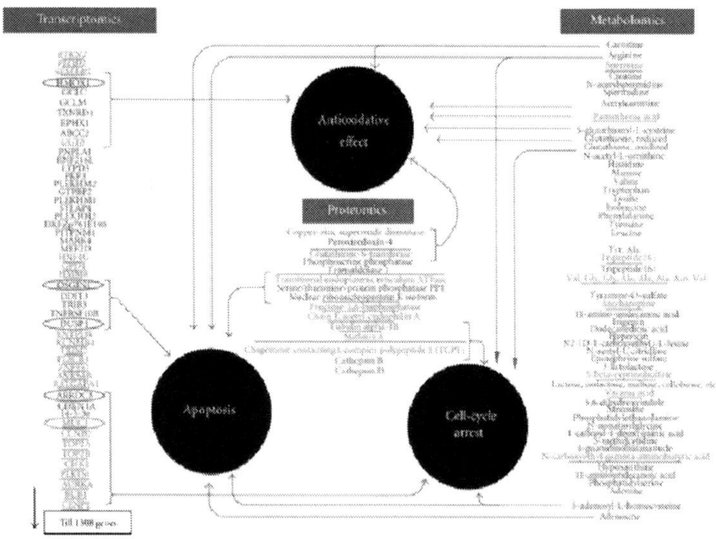

Figure 2: Foodomics identification of the genes, proteins and metabolites involved in the principal biological processes altered in HT29 colon cancer

cells after their treatment with rosemary polyphenols. In red: up-regulated; in green: down-regulated. Modified from [158] with permission from Elsevier.

However, Foodomics still needs to overcome some important gaps and limitations in order to demonstrate all its great analytical potential. Analytical strategies used in Foodomics have to face important difficulties derived, among others, from food complexity, the huge natural variability, the large number of different nutrients and bioactive food compounds, their very different concentrations, and the numerous targets with different affinities and specificities that they might have. In this context, epigenomics, transcriptomics, proteomics, and metabolomics represents powerful analytical platforms developed for the analysis of gene expression, proteins, and metabolites. However, "omics" platforms must be integrated in order to understand the biological meaning of the results on the investigated system (e.g., cell, tissue, organ) ideally through a holistic strategy as proposed by Systems Biology. Thus, Systems Biology can be defined as an integrated approach to study biological systems, at the level of cells, organs, or organisms, by measuring and integrating genomic, proteomic, and metabolic data [188]. Systems Biology approaches might encompass molecules, cells, organs, individuals, or even ecosystems, and it is regarded as an integrative approach of all information at the different levels of genomic expression (mRNA, protein, and metabolite). Data interpretation and integration when dealing with such complex systems are not straightforward and have been detected as one of the main bottlenecks. In the future, the combination of Foodomics and systems biology can provide crucial information on, for example, host-microbiome interactions, nutritional immunology, food microorganisms including pathogens resistance, farm animal production, or to fully understand postharvest phenomena through a global approach that links genetic and environmental responses and identifies the underlying biological networks. In this regard, it is expected that the new omics technologies combined with systems biology, as proposed by Foodomics, can lead postharvest research into a new era [181]. Besides, it is foreseen the emerging of other innovative approaches as, for example, green Foodomics, green systems biology [189], the human-gutome, and so forth.

ACKNOWLEDGMENTS

This work was supported by AGL2011-29857-C03-01 (Ministerio de Ciencia e Innovación, Spain) and CSD2007-00063 FUN-C-FOOD (Programa CONSOLIDER, Ministerio de Educación y Ciencia, Spain) projects.

REFERENCES

1. R. J. Mcgorrin, "One hundred years of progress in food analysis," Journal of Agricultural and Food Chemistry, vol. 57, no. 18, pp. 8076–8088, 2009.
2. A. Cifuentes, "Food analysis and foodomics," Journal of Chromatography A, vol. 1216, no. 43, pp. 7109–7110, 2009.
3. C. M. Lok and R. Son, "Application of molecularly imprinted polymers in food sample analysis—a perspective," International Food Research Journal, vol. 16, no. 2, pp. 127–140, 2009.
4. V. F. Samanidou and E. G. Karageorgou, "An overview of the use of monoliths in sample preparation and analysis of milk," Journal of Separation Science, vol. 34, pp. 2013–2025, 2011.
5. F. Wei and Y. Q. Feng, "Methods of sample preparation for determination of veterinary residues in food matrices by porous monolith microextraction-based techniques," Analytical Methods, vol. 3, no. 6, pp. 1246–1256, 2011.
6. A. Wilkowska and M. Biziuk, "Determination of pesticide residues in food matrices using the QuEChERS methodology," Food Chemistry, vol. 125, no. 3, pp. 803–812, 2011.
7. H. Z. Şenyuva and J. Gilbert, "Immunoaffinity column clean-up techniques in food analysis: a review,"Journal of Chromatography B, vol. 878, no. 2, pp. 115–132, 2010.
8. S. Balasubramanian and S. Panigrahi, "Solid-phase microextraction (SPME) techniques for quality characterization of food products: a review," Food and Bioprocess Technology, vol. 4, no. 1, pp. 1–26, 2011.
9. J. L. Tadeo, C. Sánchez-Brunete, B. Albero, and A. I. García-Valcárcel, "Application of ultrasound-assisted extraction to the

determination of contaminants in food and soil samples," Journal of Chromatography A, vol. 1217, no. 16, pp. 2415–2440, 2010.

10. M. Asensio-Ramos, J. Hernández-Borges, A. Rocco, and S. Fanali, "Food analysis: a continuous challenge for miniaturized separation techniques," Journal of Separation Science, vol. 32, no. 21, pp. 3764–3800, 2009.
11. E. Reiter, J. Zentek, and E. Razzazi, "Review on sample preparation strategies and methods used for the analysis of aflatoxins in food and feed," Molecular Nutrition and Food Research, vol. 53, no. 4, pp. 508–524, 2009.
12. M. D. Marazuela and S. Bogialli, "A review of novel strategies of sample preparation for the determination of antibacterial residues in foodstuffs using liquid chromatography-based analytical methods," Analytica Chimica Acta, vol. 645, no. 1-2, pp. 5–17, 2009.
13. A. Kumar, A. K. Malik, and Y. Picó, "Sample preparation methods for the determination of pesticides in foods using CE-UV/MS," Electrophoresis, vol. 31, no. 13, pp. 2115–2125, 2010.
14. J. Aufartová, C. Mahugo-Santana, Z. Sosa-Ferrera, J. J. Santana-Rodríguez, L. Nováková, and P. Solich, "Determination of steroid hormones in biological and environmental samples using green microextraction techniques: an overview," Analytica Chimica Acta, vol. 704, pp. 33–46, 2011.
15. A. Onal, "Overview on liquid chromatographic analysis of tetracycline residues in food matrices,"Food Chemistry, vol. 127, pp. 197–203, 2011.
16. M. Kucharska and J. Grabka, "A review of chromatographic methods for determination of synthetic food dyes," Talanta, vol. 80, no. 3, pp. 1045–1051, 2010.
17. J. Bernal, A. M. Ares, J. Pól, and S. K. Wiedmer, "Hydrophilic interaction liquid chromatography in food analysis," Journal of Chromatography A, vol. 1218, pp. 7438–7452, 2011.
18. A. L. N. van Nuijs, I. Tarcomnicu, and A. Covaci, "Application of hydrophilic interaction chromatography for the analysis of polar contaminants in food and environmental samples," Journal of Chromatography A, vol. 1218, no. 35, pp. 5964–5974, 2011.

19. D. L. Di, Y. Y. Zheng, X. F. Chen, X. Y. Huang, and S. L. Feng, "Advances in application of high-speed countercurrent chromatography in separation and purification of flavonoids," Chinese Journal of Analytical Chemistry, vol. 39, no. 2, pp. 269–275, 2011.

20. P. A. Golay, F. Giuffrida, F. Dionisi, and F. Destaillats, "Streamlined methods for the resolution and quantification of fatty acids including trans fatty acid isomers in food products by gas chromatography," Journal of AOAC International, vol. 92, no. 5, pp. 1301–1309, 2009.

21. M. Herrero, V. García-Cañas, C. Simo, and A. Cifuentes, "Recent advances in the application of capillary electromigration methods for food analysis and Foodomics," Electrophoresis, vol. 31, no. 1, pp. 205–228, 2010.

22. M. Y. Piñero, R. Bauza, and L. Arce, "Thirty years of capillary electrophoresis in food analysis laboratories: potential applications," Electrophoresis, vol. 32, no. 11, pp. 1379–1393, 2011.

23. L. Guo, B. Qiu, X. Xiao, and G. Chen, "Applications of capillary electrophoresis and microchip capillary electrophoresis for detection of genetically modified organisms," Food Science and Biotechnology, vol. 18, no. 4, pp. 823–832, 2009

24. M. Trojanowicz, A. Latoszek, and E. Pobozy, "Analysis of genetically modified food using high-performance separation methods," Analytical Letters, vol. 43, no. 10-11, pp. 1653–1679, 2010.

25. X. J. Chen, F. Q. Yang, Y. T. Wang, and S. P. Li, "CE and CEC of nucleosides and nucleotides in food materials," Electrophoresis, vol. 31, no. 13, pp. 2092–2105, 2010.

26. B. Vallejo-Cordoba and A. F. González-Córdova, "Capillary electrophoresis for the analysis of contaminants in emerging food safety issues and food traceability," Electrophoresis, vol. 31, no. 13, pp. 2154–2164, 2010.

27. G. W. Shin, H. S. Hwang, B. Chung, and G. Y. Jung, "Recent developments in CE-based detection methods for food-borne pathogens," Electrophoresis, vol. 31, no. 13, pp. 2137–2153, 2010.

28. R. Mohamed and P. A. Guy, "The pivotal role of mass spectrometry in determining the presence of chemical contaminants in food raw materials," Mass Spectrometry Reviews, vol. 30, pp. 1073–1095, 2011

29. G. Giorgi, "The 1st MS-food day conference: mass spectrometry applications in food analysis, Parma (Italy), December 2-3," Journal of Mass Spectrometry, vol. 45, p. 965, 2010.

30. H. V. Botitsi, S. D. Garbis, A. Economou, and D. F. Tsipi, "Current mass spectrometry strategies for the analysis of pesticides and their metabolites in food and water matrices," Mass Spectrometry Reviews, vol. 30, pp. 907–939, 2011.

31. D. Barceló and M. Petrovic, "Editorial: tandem MS for environmental and food analysis," Analytical and Bioanalytical Chemistry, vol. 398, no. 3, pp. 1143–1144, 2010.

32. G. Mamone, G. Picariello, S. Caira, F. Addeo, and P. Ferranti, "Analysis of food proteins and peptides by mass spectrometry-based techniques," Journal of Chromatography A, vol. 1216, no. 43, pp. 7130–7142, 2009.

33. V. García-Cañas, C. Simó, C. León, E. Ibáñez, and A. Cifuentes, "MS-based analytical methodologies to characterize genetically modified crops," Mass Spectrometry Reviews, vol. 30, no. 3, pp. 396–416, 2011.

34. M. Monagas, J. E. Quintanilla-López, C. Gómez-Cordovés, B. Bartolomé, and R. Lebrón-Aguilar, "MALDI TOF MS analysis of plant proanthocyanidins," Journal of Pharmaceutical and Biomedical Analysis, vol. 51, no. 2, pp. 358–372, 2010.

35. D. Aiello, D. De Luca, E. Gionfriddo et al., "Multistage mass spectrometry in quality, safety and origin of foods," European Journal of Mass Spectrometry, vol. 17, no. 1, pp. 1–31, 2011.

36. J. Hajslova, T. Cajka, and L. Vaclavik, "Challenging applications offered by direct analysis in real time (DART) in food-quality and safety analysis," Trends in Analytical Chemistry, vol. 30, no. 2, pp. 204–218, 2011.

37. É. Dufour, "Recent advances in the analysis of dairy product quality using methods based on the interactions of light with matter," International Journal of Dairy Technology, vol. 64, no. 2, pp. 153–165, 2011.

100. A. M. Gonzalez-Paramas, C. Santos-Buelga, M. Duenas, and S. Gonzalez-Manzano, "Analysis of flavonoids in foods and biological samples," Mini-Reviews in Medicinal Chemistry, vol. 11, pp. 1239–1255, 2011.

101. C. M. Ajila, S. K. Brar, M. Verma, R. D. Tyagi, S. Godbout, and J. R. Valero, "Extraction and analysis of polyphenols: recent trends," Critical Reviews in Biotechnology, vol. 31, pp. 227–249, 2011.

102. J. Perez-Jimenez and J. Torres, "Analysis of nonextractable phenolic compounds in foods: the current state of the art," Journal of Agricultural and Food Chemistry, vol. 59, pp. 12713–12724, 2011.

103. D. B. Rodriguez-Amaya, "Quantitative analysis, in vitro assessment of bioavailability and antioxidant activity of food carotenoids-A review," Journal of Food Composition and Analysis, vol. 23, no. 7, pp. 726–740, 2010.

104. A. Zeb and M. Murkovic, "Thin-layer chromatographic analysis of carotenoids in plant and animal samples," Journal of Planar Chromatography, vol. 23, no. 2, pp. 94–103, 2010.

105. J. Zhao, G. P. Lv, Y. W. Chen, and S. P. Li, "Advanced development in analysis of phytochemicals from medicine and food dual purposes plants used in China," Journal of Chromatography A, vol. 1218, pp. 7453–7475, 2011.

106. A. Ruiz-Rodriguez, G. Reglero, and E. Ibañez, "Recent trends in the advanced analysis of bioactive fatty acids," Journal of Pharmaceutical and Biomedical Analysis, vol. 51, no. 2, pp. 305–326, 2010

107. D. B. Clarke, "Glucosinolates, structures and analysis in food," Analytical Methods, vol. 2, no. 4, pp. 310–325, 2010.

108. J. Bernal, J. A. Mendiola, E. Ibáñez, and A. Cifuentes, "Advanced analysis of nutraceuticals," Journal of Pharmaceutical and Biomedical Analysis, vol. 55, no. 4, pp. 758–774, 2011.

109. S. A. Fade and J. A. Swift, "Qualitative research in nutrition and dietetics: data analysis issues," Journal of Human Nutrition and Dietetics, vol. 24, no. 2, pp. 106–114, 2011.

110. J. Côté, S. Caillet, G. Doyon, J. F. Sylvain, and M. Lacroix, "Analyzing cranberry bioactive compounds," Critical Reviews in Food Science and Nutrition, vol. 50, no. 9, pp. 872–888, 2010.

111. R. M. Callejón, A. M. Troncoso, and M. L. Morales, "Determination of amino acids in grape-derived products: a review," Talanta, vol. 81, no. 4-5, pp. 1143–1152, 2010

112. V. L. Campo, D. F. Kawano, D. B. D. Silva, and I. Carvalho, "Carrageenans: biological properties, chemical modifications and structural analysis—a review," Carbohydrate Polymers, vol. 77, no. 2, pp. 167–180, 2009.

113. M. R. Kasaai, "Various methods for determination of the degree of N-acetylation of chitin and chitosan: a review," Journal of Agricultural and Food Chemistry, vol. 57, no. 5, pp. 1667–1676, 2009.

114. T. Korenaga, "Stable isotope ratio analyses of light elements and their application to verifying geographical origin of foods," Journal of the Food Hygienic Society of Japan, vol. 51, no. 5, pp. 197–204, 2010.

115. L. E. McDonald, J. Brett, Kelton, D. Majowicz S. E, K. Snedeker, and J. M. Sargeant, "A systematic review and meta-analysis of the effects of pasteurization on milk vitamins, and evidence for raw milk consumption and other health-related outcomes," Journal of Food Protection, vol. 74, pp. 1814–1832, 2011.

116. C. Bicchi, C. Cagliero, and P. Rubiolo, "New trends in the analysis of the volatile fraction of matrices of vegetable origin: a short overview. A review," Flavour and Fragrance Journal, vol. 26, pp. 321–325, 2011.

117. M. Herrero, C. Simó, V. García-Cañas, S. Fanali, and A. Cifuentes, "Chiral capillary electrophoresis in food analysis," Electrophoresis, vol. 31, no. 13, pp. 2106–2114, 2010.

118. N. Perrot, I. C. Trelea, C. Baudrit, G. Trystram, and P. Bourgine, "Modelling and analysis of complex food systems: state of the art and new trends," Trends in Food Science and Technology, vol. 22, no. 6, pp. 304–314, 2011.

119. M. H. Tunick, "Food texture analysis in the 21st century," Journal of Agricultural and Food Chemistry, vol. 59, no. 5, pp. 1477–1480, 2011.

120. R. E. Levin, Rapid Detection and Characterization of Foodborne Pathogens by Molecular Techniques, CRC Press, 2009.

121. D. Liu, Molecular Detection of Foodborne Pathogens, CRC Press, 2009.

122. B. Popping, C. Diaz-Amigo, and K. Hoenicke, Molecular Biological and Immunological Techniques and Applications for Food Chemists, John Wiley & Sons, 2009.

123. J. Zweigenbaum, Mass Spectrometry in Food Safety, Humana Press, 2011.

124. S. Otles, Handbook of Food Analysis Instruments, CRC Press, 2009.

125. L. M. L. Nollet and F. Toldra, Handbook of Dairy Foods Analysis, CRC Press, 2009.

126. G. S. Bennet, Food Identity Preservation and Traceability: Safer Grains, CRC Press, 2009.

127. S. N. Mahindru, Food Contaminants-Origin, Propagation and Analysis, APH Publishing Corporation, 2009.

128. J. Wang, J. D. MacNeil, and J. F. Kay, Chemical Analysis of Antibiotic Residues in Food, Interscience Series on Mass Spectrometry, John Wiley & Sons, 2011.

129. M. Rychlik, Fortified Foods with Vitamins: Analytical Concepts to Assure Better and Safer Products, Wiley-VCH, 2011.

130. L. E. Rodriguez-Saona and M. E. Allendorf, "Use of FTIR for Rapid authentication and detection of adulteration of food," in Annual Review of Food Science and Technology, vol. 2, pp. 467–483, 2011.

131. J. van Duynhoven, A. Voda, M. Witek, and H. Van As, "Time-domain NMR applied to food products," in Annual Reports on NMR Spectroscopy, G. A. Webb, Ed., vol. 69, pp. 145–197, 2010.

132. T. Lavecchia, A. Tibuzzi, and M. T. Giardi, "Biosensors for functional food safety and analysis," inBio-Farms for Nutraceuticals: Functional Food and Safety Control by Biosensors, M. T. Giardi, G. Rea, and B. Berra, Eds., Landes Biosciences, 2010.

133. L. Mondello, F. M. Lancas, and P. Q. Tranchida, "Food analysis," Journal of Separation Science, vol. 32, no. 21, pp. 3573–3855, 2009.

134. A. Cifuentes, "Advanced separation methods in food analysis," Journal of Chromatography A, vol. 1216, pp. 7109–7358, 2009.

135. A. Paschke and F. Ulberth, "Editorial: allergens in foods," Analytical and Bioanalytical Chemistry, vol. 395, no. 1, pp. 15–16, 2009.

136. B. Chankvetadze and A. Cifuentes, "Natural bioactive compounds and Nutrigenomics," Journal of Pharmaceutical and Biomedical Analysis, vol. 51, no. 2, pp. 289–496, 2010.

137. A. Cifuentes, "Editorial food science, foodomics and capillary electromigration methods,"Electrophoresis, vol. 31, no. 13, pp. 2091–2296, 2010.

138. Cifuentes, A. Dugo P, and S. Fanali, "Advanced food analysis," Journal of Chromatography A, vol. 1218, pp. 7385–7738, 2011.

139. E. Ibañez and A. Cifuentes, "New analytical techniques in food science," Critical Reviews in Food Science and Nutrition, vol. 41, no. 6, pp. 413–450, 2001.

140. J. Zweigenbaum, "Contaminants in our food? How to detect untargeted compounds and determine their identity," Agro Food Industry Hi-Tech, vol. 22, pp. 4–6, 2011.

141. L. Kantiani, M. Llorca, J. Sanchís, M. Farré, and D. Barceló, "Emerging food contaminants: a review,"Analytical and Bioanalytical Chemistry, vol. 398, no. 6, pp. 2413–2427, 2010.

142. K. Lancova, R. Dip, J. P. Antignac, B. L. Bizec, C. T. Elliott, and H. Naegeli, "Detection of hazardous food contaminants by transcriptomics fingerprinting," Trends in Analytical Chemistry, vol. 30, no. 2, pp. 181–191, 2011.

143. M. Severgnini, P. Cremonesi, C. Consolandi, G. De Bellis, and B. Castiglioni, "Advances in DNA Microarray Technology for the Detection of Foodborne Pathogens," Food and Bioprocess Technology, vol. 4, no. 6, pp. 936–953, 2011.

144. H. P. Dwivedi and L. A. Jaykus, "Detection of pathogens in foods: the current state-of-the-art and future directions," Critical Reviews in Microbiology, vol. 37, no. 1, pp. 40–63, 2011.

145. R. M. González-Rodríguez, R. Rial-Otero, B. Cancho-Grande, C. Gonzalez-Barreiro, and J. Simal-Gándara, "A review on the fate of pesticides during the processes within the food-production Chain,"Critical Reviews in Food Science and Nutrition, vol. 51, no. 2, pp. 99–114, 2011.

146. K. Campbell, N. Vilariño, L. M. Botana, and C. T. Elliott, "A European perspective on progress in moving away from the mouse bioassay for marine-toxin analysis," Trends in Analytical Chemistry, vol. 30, no. 2, pp. 239–253, 2011.
147. A. Marques, H. M. Lourenço, M. L. Nunes et al., "New tools to assess toxicity, bioaccessibility and uptake of chemical contaminants in meat and seafood," Food Research International, vol. 44, no. 2, pp. 510–522, 2011.
148. P. Coleman, "Microbial risk analysis of foods," Risk Analysis, vol. 30, pp. 866–871, 2010.
149. S. Hoffmann, "Ensuring food safety around the globe: the many roles of risk analysis from risk ranking to microbial risk assessment: introduction to special series," Risk Analysis, vol. 30, no. 5, pp. 711–714, 2010.
150. M. J. Tijhuis, N. de Jong, M. V. Pohjola et al., "State of the art in benefit-risk analysis: food and nutrition," Food and Chemical Toxicology, vol. 50, pp. 5–25, 2012.
151. L. Jacxsens, P. A. Luning, W. J. Marcelis et al., "Tools for the performance assessment and improvement of food safety management systems," Trends in Food Science and Technology, vol. 22, pp. S80–S89, 2011.
152. B. A. Magnuson, T. S. Jonaitis, and J. W. Card, "A brief review of the occurrence, use, and safety of food-related nanomaterials," Journal of Food Science, vol. 76, no. 6, pp. R126–R133, 2011.
153. Y. G. Vlasov, Y. E. Ermolenko, A. V. Legin, A. M. Rudnitskaya, and V. V. Kolodnikov, "Chemical sensors and their systems," Journal of Analytical Chemistry, vol. 65, no. 9, pp. 880–898, 2010.
154. M. G. Valdés, A. C. V. González, J. A. G. Calzón, and M. E. Díaz-García, "Analytical nanotechnology for food analysis," Microchimica Acta, vol. 166, no. 1-2, pp. 1–19, 2009.
155. B. Perez-Lopez and A. Merkoci, "Nanomaterials based biosensors for food analysis applications,"Trends in Food Science & Technolog, vol. 22, pp. 625–639, 2011.
156. S. R. Raz and W. Haasnoot, "Multiplex bioanalytical methods for food and environmental monitoring," Trends in Analytical Chemistry, vol. 30, pp. 1526–1537, 2011.

157. Y. T. Atalay, S. Vermeir, D. Witters et al., "Microfluidic analytical systems for food analysis," Trends in Food Science and Technology, vol. 22, no. 7, pp. 386–404, 2011.

158. C. Ibáñez, A. Valdés, V. García-Cañas et al., "Global Foodomics strategy to investigate the health benefits of dietary constituents," Journal of Chromatography A, vol. 1248, pp. 139–153, 2012.

159. M. Herrero, E. Ibáñez, A. Cifuentes, and J. Bernal, "Multidimensional chromatography in food analysis," Journal of Chromatography A, vol. 1216, no. 43, pp. 7110–7129, 2009.

160. P. Q. Tranchida, P. Dugo, G. Dugo, and L. Mondello, "Chapter 10 Analysis of Food Constituents,"Comprehensive Analytical Chemistry, vol. 55, pp. 215–241, 2009.

161. J. E. Welke and C. A. Zini, "Comprehensive two-dimensional gas chromatography for analysis of volatile compounds in foods and beverages," Journal of the Brazilian Chemical Society, vol. 22, no. 4, pp. 609–622, 2011.

162. S. Armenta, S. Garrigues, and M. de la Guardia, "Green Analytical Chemistry," Trends in Analytical Chemistry, vol. 27, no. 6, pp. 497–511, 2008.

A

Results for a General Estimation Equation model describing the income from the sale of chickens under three different treatments from August 2004 until June 2005 in Myanmar. TR1, I-2 Newcastle disease vaccination; TR2, placebo treatment only; TR3, altered management of chick rearing; details of treatment groups are provided in the text.

Explanatory variables	Difference between mean income from sale of chickens (relative to reference category) (Kyat)	95% CI		p
TR1	Reference category			
TR2	787.53	9.86	1565.20	0.05
TR3	2513.06	1853.30	3172.83	<0.001
August 2004	Reference category			
September 2004	-3004.88	-4631.18	-1378.58	<0.001
October 2004	-3891.92	-6258.92	-1524.92	<0.001
November 2004	-4890.82	-6265.54	-3516.10	<0.001
December 2004	-4888.37	-6877.79	-2898.95	<0.001
January 2005	1507.58	-185.49	3200.65	0.08
February 2005	-2984.49	-4585.95	-1383.03	<0.001
March 2005	-4355.45	-5588.49	-3122.41	<0.001
April 2005	-1109.12	-2787.44	569.21	0.20
May 2005	-5775.45	-8991.01	-4559.90	<0.001
June 2005	-5418.21	-8359.74	-2476.68	<0.001
Constant	9232.92	165.64	4867.10	<0.001

B

Figure 2: Methods to demonstrate the increased income from the sale of chickens by adopting village chicken management interventions in Myanmar. A) Results from a General Estimation Equation Model estimating income from the sale of chickens (Henning et al. 2009) and B)Extension cartoon showing the number of additional birds produced and additional money earned from the sale of chickens.

A serological survey combined with an on-farm recording of flock mortalities indicated that Newcastle disease outbreaks predominately occur in the months of March and April (Figure 3A1 and Figure 3A2) (Henning et al. 2008). An annual chicken flock survival circle was developed showing individual months and the number of birds being alive or succumbing from disease in each month (Figure 3B1). The months when Newcastle disease vaccination using I-2 vaccine should be conducted to prevent these outbreaks are indicated with letters 'I-2'.

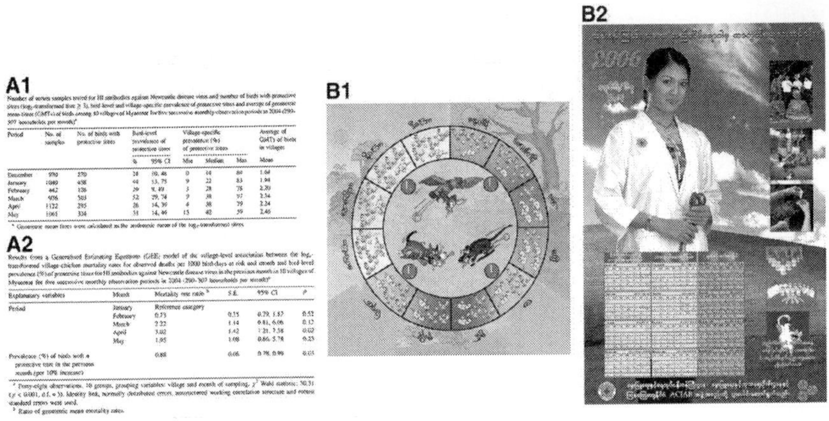

Figure 3: Methods to describe the seasonal Newcastle disease occurrence in Myanmar. A1) Bird and village level sero-prevalence of Newcastle disease antibodies over 6 months in village chickens (Henning et al. 2008) and A2) Results from a General Estimation Equation Model estimating village chicken mortality rates per month B1) Extension cartoons showing an annual cycle of mortality including the number of birds being alive or dying in a particular month and indicating months when Newcastle disease vaccination (I2) is required and B2) Newcastle disease vaccination calendar with a Myanmar actress posing as a veterinarian and the months when vaccination is required being highlighted (in red) (Henning et al. 2009).

Farmer workshops were the major tool to deliver the extension messages on improving village chicken health and on the use of the intervention methods. These workshops were conducted in common spiritual or communal places (e.g. in monasteries, in the house of the village head person, in schools or in village offices) from June until September 2007 in 47 villages in three different divisions of Myanmar (Yangon, Bago, Mandalay Divisions). Women and children as the main poultry keepers were the main target in the workshops. Training of trainers resulted in training of 14 field veterinarians and 47 village head people to deliver the extension messages on village chicken health and improved chick husbandry. Overall a total of 166 workshops were conducted with 3,969 trainees attending (representing about 50% of the 9,157 village households in the 47 villages). Farmers who had positive experiences with the interventions in the field studies were invited to report on their experiences. Farmers could purchase the extension equipment (bamboo coops, creep feeders, chick starter feed) used in the intervention studies. During the trainings sessions a total of 1750 complete sets of extension equipment for improved chick management were sold to farmers. Workshop participation was encouraged through lotteries in which participants could win some of the extension equipment. Village personnel who had significant association with poultry issues were identified to conduct these workshops in the future.

Extension messages comprising cartoons, diagrams and images were incorporated into the following extension materials:

- Flip charts with extension messages painted on strong, water proof canvas were used during farmer workshops. These charts presented the extension messages in sequence as a 'theatre of messages', they were 'hands-on' material and provided an entertainment to the audience. After the workshops the flip charts were further displayed in the schools of the villages.
- Booklets provided 'take home' messages and were handed out at the end of workshops. In addition to the cartoons, these booklets contained simple written statements and provided a readily available reference for the farmer.• Vaccination calendars showing a famous Myanmar actress posing as a LBVD veterinarian (to encourage the display of the calendar in the farmers' homes) and highlighting the months for I-2 vaccination were developed (Figure 3B2). This vaccination calendar was very popular and several editions had to be produced.
- Posters with extension messages were shown in central places such as major cross-roads or central locations, in or near the township veterinary offices, the house of the village head man or near shops or village markets.

DISCUSSION

Extension can be understood as providing a link between research and farmers. However, this only works if an appropriate method of communication is used. This project aimed to directly 'translate' scientific research outcomes in an understandable manner to the local village chicken owning community in Myanmar, and to use appropriate communication strategies. For example large, single-page calendars are often used as decoration in farm houses in Myanmar. We attracted farmers' interest in the vaccination program by incorporating a photo of a famous Myanmar actress posing as a veterinarian in a calendar. Vaccination against Newcastle disease using I-2 vaccine is usually conducted three times per year, although the natural brooding in village flocks and the thereby continued introduction of susceptible young birds has to be considered in the vaccination scheme (Spradbrow 2014). Therefore LBVD of Myanmar recommended village chicken vaccinations four times per year and we incorporated this frequency in our extension messages.

General extension material on clinical signs of Newcastle disease and vaccination practices have been produced previously (Ahlers et al. 2009), but we developed extension messages that reflected specific published results obtained from our own village poultry research conducted in Myanmar. When preparing extension messages using cartoons or drawings, country-specific characteristics, religious and traditional beliefs have to be considered (Alders and Bagnol 2007). Extension messages developed for situations in other countries often do not work and have to be modified (e.g. poultry markets differ substantially between developed and less developed South East Asian countries). Furthermore pictures or cartoons should provide only single messages or limited information in local surroundings and ideally they should be colourful and entertaining. As with questionnaire development, pilot-testing of extension pictures and materials is an imperative. An important request made by farmers during the pilot testing of our images was that 'bubbles of souls moving on' should be incorporated into images showing dead birds, describing the goal of the Buddhist path that death is not the end of life, but the end of the body and the spirit will still remain.

The major tool to deliver our extension messages were village meetings. Other means of extension that we considered using in our project included radio, television and marionette theatre. In contrast to most African societies, where many local radio stations exist, radio transmission is not so common in Myanmar. Only government owned and controlled radio stations existed at the time of the project and Myanmar farmers were not accustomed to listening to radio programs. Therefore we evaluated that the use of radio as an extension method was not valuable. In contrast, television (TV) programs (two Myanmar government owned TV stations existed at the time) are often watched by village people in local tea shops or, if available, in the house of the TV owner in the village (although most villages had no or infrequent power supply) and could have been a useful medium for our extension messages. However, because 'foreign' involvement in TV advertisement at the time of the project was discouraged by government officials, we decided not to produce a TV advertisement on village chicken health.

Marionette theatre has a long tradition in Myanmar and Myanmar people have a great respect for an expert puppeteer. A puppet play would have been a useful tool to deliver a message on improving village chicken health to a wider audience as travelling puppeteers often visit even remote villages. In particular children who are often looking after chickens in a village household could be reached via this method. However, within the time frame of the project it was not possible to deliver a puppet play on village chicken health.

Communication, defined as the sharing of ideas and information, forms a large part of the extension worker's job. Passing on ideas, advice and information to influence the decisions of farmers is a bi-directional process in which the extension worker has to listen carefully to the farmers' needs. Farmers might be reluctant to engage in innovations as they might be perceived as too risky (Berdegue 1992). Thus it is necessary to explore farmers' perceptions in a close collaboration between different disciplines such as veterinary science, social science, anthropology and agricultural science (Forno 1999; Swanson et al. 1997). We believe that epidemiological survey methodology and participatory approaches are important not only to explore current perceptions, attitudes and practices, but also traditional beliefs. Therefore a combination of quantitative and qualitative survey methods is essential to understand what extension message might work or won't work for a farmer and/or for a village or community under the current conditions.

The spoken word is a key communication tool, but, whether the person delivering the extension message is speaking to a large village meeting or is discussing a problem in the field with a group of farmers, its impact and effectiveness can be greatly increased by the use of suitable audio-visual aids (Oakley and Garforth 1985). Sophisticated audio-visual aids might require electricity and complex equipment such as projectors or televisions, which are often not available in villages of low income countries. Simple aids that can be made locally such as flip charts have several advantages over audio-visual equipment: they do not require a power source, they do not cost much to produce and they can be made to suit the precise needs of the farmers.

Often epidemiological research studies provide a variety of outcomes, but identifying the most important messages to be communicated to farmers can be a challenge. Interventions will only be acceptable by the farmer, if they satisfy the farmer's need; they have to be technically, economically and ecologically sound. Some needs perceived by farmers might be inappropriate (e.g. upgrading domestic village chicken breeds with exotic breeds that perform poorly in the same scavenging environment) and this has to be carefully discussed with farmers. Also commercial operations will have different constraints than small-scale producers, and preventive messages might not be applicable to both enterprise types in the same way (e.g. fencing-off areas for scavenging village chickens to improve biosecurity might not be feasible for small-scale family poultry farmers).

Sumberg (2005) indicated that research outputs that are not used by farmers are, by definition, not relevant to farmers, and the reasons for this 'irrelevance' must be further examined. Hence the adoption process is the procedure in which farmers choose or choose not to use new ideas (or interventions) on their farms (Luukkainen 2012). It is a process that involves gaining new information (usually provided by the extension advisor), a development of interest by the farmer in which he/she changes their attitudes towards the idea, an evaluation and decision making process in which the farmers decides to use or not use the idea, a trial and implementation phase in which the idea is tested, and finally the adoption and confirmation phase, in which the farmer decides if the new idea is preferable over the old methods (Luukkainen 2012). What drives farmers to change or not to change farming practices depends on factors associated with implementation of the idea (e.g. costs and saved expenses, marketing opportunities for the additional product produced, time commitment for implementing the new idea), characteristics of the individual (e.g. age, education level, occupation), community and ecological variables (e.g. disease situation in the village or region, actions that neighbours or relatives implement related to the new idea) and larger social, administrative and political variables (e.g. government, pricing and trade policies) (Leeuwis 2004). All these variables, including the varying adoption outcomes (e.g. short-term, long-term, and perhaps unintended adoption) should be explored in epidemiological-sociological investigations to evaluate the success of the adoption process. Although farmers were eager to adopt the simple interventions in our project, the long-term adoption could not be accessed within the timeframe of the study. The availability of the intervention equipment was probably the most limiting factor for the successful adoption of our promoted interventions. This included the continued production and availability of the I-2 vaccine manufactured by LBVD and the availability of small packages of chick starter feed for village chicken farmers.

Several economic factors are important to both the scope of the activity and the perception of extension assistance to farmers. The first is the profit opportunity available to farmers who improve productivity by applying new technologies brought to their attention (Swanson et al. 1997). Farmer response to new agricultural innovations is directly related to financial advantage from applying such recommendations. We had shown that management changes applied in the intervention study resulted in increased productivity and greater income for the farmer (Henning et al. 2009, 2013) and this higher income was directly shown in extension messages. A second factor is the existing marketing system. Marketing represents the process by which an individual farm or family produce is transferred from the farm to become a value-laden commodity. Mostly live birds are sold in Myanmar for cash to middle men or they are sold by farmers directly in local village markets. Therefore including the marketing of birds in the extension messages was essential. The third factor relates to the general economic environment in which a nation's agriculture finds itself. The more prosperous agriculture is, the less interest there may be in seeking assistance from an extension service. Myanmar is an agriculture dominated economy and the income per capita is one of the lowest in the region. Hence we experienced very strong interest by village chicken farmers to learn about methods to improve village chicken productivity.

In addition to the factors described previously, we identified a number of specific variables that can influence the success or otherwise of veterinary extension programs in small-scale agricultural settings.

Role of Women

Specific livestock species are often managed by a specific gender in a farmer's family. In the baseline survey we identified that in 78%

of our study households, women were the people working with chickens (Henning et al. 2007). Therefore we actively encouraged the participation of women in the farmer meetings.

Level of Illiteracy

Extension services must operate at the local community level and must address the levels of literacy and education as they affect extension directly (Bagnol 2012). As most village chicken farmers in our project areas were illiterate we used pictures and cartoons rather than complicated written explanations in our extension messages.

Suitability of Interventions to Farming Conditions

Sufficient knowledge and understanding of the farming system is essential to make it sustainable. We used household surveys to identify farmers' circumstances in the target area (Henning et al. 2007), then new technologies/intervention strategies were planned, designed and followed by on-farm testing for a period of one year (Henning et al. 2009). In transferring new technology, it is important to understand a farmer's own evaluation of its performance during early adoption. Feedback from farmers on the research methodology used and on equipment promoted was collected and recommendations given by farmers during pilot-testing and extension workshops were followed up. For instance creep feeders for chick feeding were modified based on the advice given by farmers early in the project.

Farmers' Resources to Adopt Interventions

A serious limitation in rural development is the lack of available capital on which to develop a base of economic activity (Swanson et al. 1997). For successful adoption, interventions promoted have to be affordable for farmers. In our project most village chicken farmers could afford the equipment promoted. However, in the beginning of our extension program farmers were able to purchase chick management equipment from the project at subsidized prices to encourage its use. If larger investments are required to adopt interventions, credit opportunities for farmers need to be explored.

ACKNOWLEDGEMENTS

We thank all Myanmar farmers who made this research into village chicken health and production and the development of extension messages possible. The project was supported by Australian Centre for International Agricultural Research aid program (project number AH/2002/042).

REFERENCES

1. Agbarevo MNB (2013) Farmers' perception of effectiveness of agricultural extension delivery in cross-river state, Nigeria. IOSR-JAVS 2(6):01-07
2. Ahlers C, Alders RG, Bagnol B, Cambaza AB, Harun M, Mgomezulu R, Msami H, Pym B, Wegener P, Wethli E, Young M (2009) Improving Village Chicken Production: A Manual for Field Workers and Trainers. Australian Centre for International Agricultural Research, Canberra, Australia.
3. Alders RG, Bagnol B (2007) Effective communication: the key to efficient HPAI prevention and control. Worlds Poult Sci J 63:139-147 Publisher Full Text
4. Alders RG, Bagnol B, Young MP (2010) Technically sound and sustainable Newcastle disease control in village chickens:lessons learnt over fifteen years. Worlds Poult Sci J 66:433-440
5. Bagnol B (2012) Good Practices for Family Poultry Production. Advocate Gender Issues: A Sustainable Way to Control Newcastle Disease in Village Chickens. International Network for Family Poultry Development. GPFPP Note No 03. Food and Agriculture Organization of the United Nations, Rome, Italy. Available from http://www.fao.org/docrep/018/aq630e/aq630e.pdf *webcite* Accessed 15 Jun 2014
6. Berdegue JA (1992) Challenges of Farming Systems Research and Extension. In: Keynote Presentation at the 12th Annual Association for Farming Systems Research – Extension Symposium. East Lansing, USA: Michigan State University.
7. Chemnitz C, Becheva S (2014) The Meat Atlas. Heinrich Böll Foundation, Berlin, Germany & Friends of the Earth Europe,

Brussels, Belgium. Available fromhttp://www.foeeurope.org/sites/default/files/publications/foee_hbf_meatatlas_jan2014.pdf Accessed 15 Jun 2014

8. Forno DA (1999) Sustainable Development Starts with Agriculture. In: Sustainable Agriculture Solutions. London, UK: Novello Press Ltd.
9. Henning J, Morton J, Pym R, Hla T, Sunn K, Meers J (2013) Economic analysis of interventions to improve village chicken production in Myanmar. Prev Vet Med 110:525-540
10. Henning J, Morton J, Pym R, Hla T, Meers J (2009) Evaluation of strategies to improve village chicken production - controlled field trials to assess effects of Newcastle disease vaccination and altered chick rearing in Myanmar. Prev Vet Med 90:17-30
11. Henning J, Morton J, Hla T, Meers J (2008) Mortality rates adjusted for unobserved deaths and associations with Newcastle disease virus serology among unvaccinated village chickens in Myanmar. Prev Vet Med 85:241-252
12. Henning J, Pym R, Hla T, Kyaw N, Meers J (2007) Village chicken production in Myanmar–purpose, magnitude and major constraints. Worlds Poult Sci J 63:309-323
13. Leeuwis C (2004) Communication for Rural Innovation: Rethinking Agricultural Extension. Blackwell Science, UK.
14. Luukkainen J (2012) A Comparison of Extension Methods Used by Different Agricultural Extension Service Providers in Nyandarua County, Kenya. HAMK University of Applied Sciences, Forssa, Finland.
15. McCrindle CME (1995) The Practice of Veterinary Extension: What is it and Who should be doing it. Proceedings of the National Forum on State Veterinary Matters, Pilanesberg, South Africa.
16. Oakley P, Garforth C (1985) Guide to Extension Training. Agriculture Extension and Rural Development Centre, School of Education, University of Reading, UK. Food and Agriculture Organization of the United Nations, Rome, Italy. Available from http://www.fao.org/docrep/t0060e/t0060e00.HTM *webcite* Accessed 15 Jun 2014
17. Spradbrow P (2014) Thermotolerant Newcastle disease vaccines for use in flocks of family chickens. CAB Reviews 9:No. 011

18. Sumberg J (2005) Systems of innovation theory and the changing architecture of agricultural research in Africa. Food Policy 30:21-41

19. Swanson BE, Bentz PB, Sofranko AJ (1997) Improving Agricultural Extension: A Reference Manual. Food and Agriculture Organization of the United Nations, Rome, Italy. Available from http://www.fao.org/docrep/w5830e/w5830e00.HTM Accessed 15 Jun 2014

20. Swanson BE, Claar JB (1984) The History and Development of Agricultural Extension. In: Swanson BE (ed) Agricultural Extension. A Reference Manual, Rome, Italy: Food and Agriculture Organization of the United Nations. pp 1-19

Chapter 6

Pastoralist Livelihoods and Wildlife Revenues in East Africa: A Case for Coexistence?

Katherine M Homewood[1], Pippa Chenevix Trench[2], and Daniel Brockington[3]

[1]Anthropology, University College London, Gower Street, London, WC1E 6BT, UK

[2]Markets Trade and Institutions Division, International Food Policy Research Institute, 2033 K Street, NW, Washington, DC, 20016, USA

[3]Institute for Development Policy and Management, School of Environment and Development, the University of Manchester, Oxford Road, Manchester, M13 9PL, UK

ABSTRACT

East African arid and semi-arid lands are home to many of the world's pastoralists and most spectacular savanna wildlife populations, attracting substantial conservation and tourism revenues. Yet these peoples are among the poorest (and most affected by extreme climatic events), and the wildlife is in unsustainable decline. National governments, international donors and conservation agencies favour win-win solutions through conservation with development. Maasailand is a hotspot of conservation, poverty and new initiatives to redistribute tourist income. We outline pastoralist livelihoods and how these are changing, then summarise status and trends of wildlife populations, tourism revenues, and conservation and development initiatives in East Africa and Maasailand. We ask to what extent wildlife revenues contribute to pastoralist livelihoods and whether this translates into a robust basis for coexistence. To put in context the role and importance of wildlife- and tourism-based activities, we outline findings from a multi-site study of Maasai livelihoods. Livestock contribute half or more of the mean annual income in all sites, with off-farm work and farming ranking second and third, respectively, except in Mara, where wildlife-based income contributes around 20% income across all wealth categories. In most sites, significant areas have been set aside for conservation and tourism, but wildlife contributes <5% income to a small proportion of households at most. Few wildlife-derived benefits flow to pastoralists, while conservation restrictions constrain production and coping strategies, undermining potential for coexistence. In exceptional circumstances, significant wildlife revenue may reach households, but full social and ecological implications of associated conservancy agreements remain unclear.

BACKGROUND

East African arid and semi-arid lands (ASAL) are home to many of the world's pastoralists and most spectacular savanna wildlife populations, attracting substantial conservation and tourism revenues. Yet these peoples are among the poorest (and most affected by extreme climatic events), and wildlife is in unsustainable decline. National governments, international donors and conservation agencies favour win-win solutions through conservation with development. On the face of it, conditions appear favourable for pastoralists and wildlife to coexist in sustainable ways, with tourist revenues compensating for the livelihood impacts of any conservation restrictions. National governments, bilateral donors and international conservation non-governmental organisations (NGOs) envisage such initiatives as partly or wholly replacing pastoralist livestock production, which has often been assumed to be environmentally damaging, with what are seen as more ecologically and economically sustainable wildlife-based forms of land use. Although there have been problems with transparency and accountability, in Maasailand there is a rapid proliferation of initiatives pursuing conservation with development, and these are fast evolving new institutions managing finances and monitoring enforcement and governance.

To put the contribution of wildlife tourism to poverty reduction and local livelihoods in context, we first outline pastoralist livelihoods and how these are changing, then summarise status and trends of wildlife populations, tourism revenues, and conservation and development initiatives in East Africa and Maasailand. This paper then asks to what extent wildlife revenues contribute to pastoralist livelihoods and whether this translates into a robust basis for coexistence. To do this, we look first at the role livestock and other activities play in rural Maasai household economies and in livelihoods overall. Recent analyses have cast households into four economic groups according to their position on the two axes of livestock holdings and cash income (McPeak et al. 2012). The studies we summarise here sought to capture a further dimension of livelihood strategies: the patterns of household-level diversification into different livelihood activities. Individual studies are reported in (Homewood et al. 2009). These studies sought a balanced and contextualised view of the contribution of wildlife conservation to local livelihoods in rural Maasailand by asking what were people doing, how well they did, what factors shaped choice of income-earning activities and what factors influenced how well they did. They explored whether identifiable livelihood strategies emerged in relation to particular predictors, and within any such strategies, looked for the significance of wildlife-related activities and income. Taking a comparative approach across Kenyan and Tanzanian study sites, these data allow evaluation of conservation and poverty reduction policies and practices.

We argue that rural Maasai land use decisions do not support national- and international-level assumptions about the benefits of wildlife and tourism, nor about a relatively lower economic importance of livestock production. However unequally distributed, and however insufficient in themselves to sustain families, livestock emerge as school or sick people attending hospital. Some pastoralists are able to intensify, using supplements, veterinary inputs, fodder crops, tighter integration of farming and livestock, 'improved' high-yielding (but drought- and disease-prone) breeds and more market-oriented models of production. For example, better-off herd owners may buy up poorer families' young male calves and vaccinate them, enhancing survival and rearing a much higher proportion to commercially valuable size (Homewood ET al.2006).

What is the role of wildlife-based incomes in this complex tapestry of pastoralist livelihoods? Diversification and intensification mean increasingly sedentary life ways that allow access to farm work, jobs, markets, schools and hospitals, but impact adversely on livestock mobility, performance and production. Two crosscutting trends make pastoral production harder: land fragmentation (and land alienation) constraining mobility, and climate change, which is expected to make mobility even more crucial (Galvin 2009; Davies and Nori 2008). Alone among the many potential pathways of diversification outlined above, wildlife-based activities commonly promote more open rangelands theoretically consistent with mobile pastoralist production strategies. With wild grazers and domestic livestock depending on the same key resources, and pastoralists using mobility and migration strategies that closely parallel those whereby wildlife make the most of patchy and unpredictable grazing and water, there are synergies to be found. However, the overlap also potentially makes for competition and tradeoffs between wildlife and pastoralist land uses.

Tensions around tradeoffs are intensified by policy. Kenyan and Tanzanian governments see pastoralist livestock management (mobile transhumance on unfenced, unmodified rangelands) as unproductive and environmentally damaging e.g. (URT 1997; MoLF. 2006). Pastoral migration to south Tanzania is perceived, without good data, to be driven by pastoralists' own degradation of their rangelands (Brockington 2006). Regional and district governments impose draconian confiscations of cattle and fines, constraining pastoral activities while benefiting from their productivity. In contrast to the generally negative official take on mobile pastoralism, wildlife tourism is portrayed by these same agencies as an environmentally and economically sustainable means for pastoral groups to diversify, generate revenues and improve well-being. These schemes almost invariably aim for wildlife activities to replace pastoralism through a part of its range, and thus, for the two to coexist alongside one another, rather than overlapping fully within the same space (as was originally intended to be the case for multiple land use areas like the Ngoro Ngoro Conservation Area, and as is envisaged for buffer zones around protected areas).

Wildlife, Tourism, Conservation and Development in East Africa

Conservation is big business in East Africa. Tourism is regularly among the top three contributors to GDP and to foreign exchange earnings in Kenya, accounting for USD 884 million in 2010 Kshs 73.7 billion; (Ministry of Tourism2010). Despite the global financial crisis, Tanzania earned USD 1.16 billion from tourism in 2009 (The Tanzania National Website 2009). In both countries, tourists are strongly drawn by the appeal of wildlife alongside other attractions. Conservationists see tourists' dollars as one of the principal means to generate meaningful income for the rural poor (Roe et al. in press). Government policies (URT 20022005; UNDP/UNEP/IIED/IUCN/WRI 2005), conservation NGO projects (AWF 2005), entrepreneurial initiatives (Nelson 2004; Lewa WildlifeConservancy 2010) and research publications (Pearce and Moran 1994; Hutton et al. 2005) all promote wildlife-based tourism. Maasailand, the region of Kenya and Tanzania dominated by Maa-speaking pastoralists, is a hotspot of conservation, poverty and new initiatives to redistribute tourist income and is a good place to explore dynamics and distribution of revenues.

In Kenya, Maasailand and other pastoral areas represent among the fastest growing tourism destinations (33% growth in bed-nights 2004 to 2005; Ministry of Tourism and Wildlife 2006). In Tanzania In 2009, 16 National Parks earned USD 43.8 million; Ngorongoro Conservation Area, USD 22.6 million; and tourist hunting, USD 14.9 million (The Tanzania National Website 2009). In both Kenya and Tanzania, the highest earning protected areas are situated within, and effectively excised from, Maasailand (Figure 1), as is a high proportion of the two countries' conservation estate overall. Parts of Kenyan Maasailand have shown rapid economic growth driven by wildlife conservation, rising domestic and export markets for crops and rising land values (Norton-Griffiths and Said 2010).

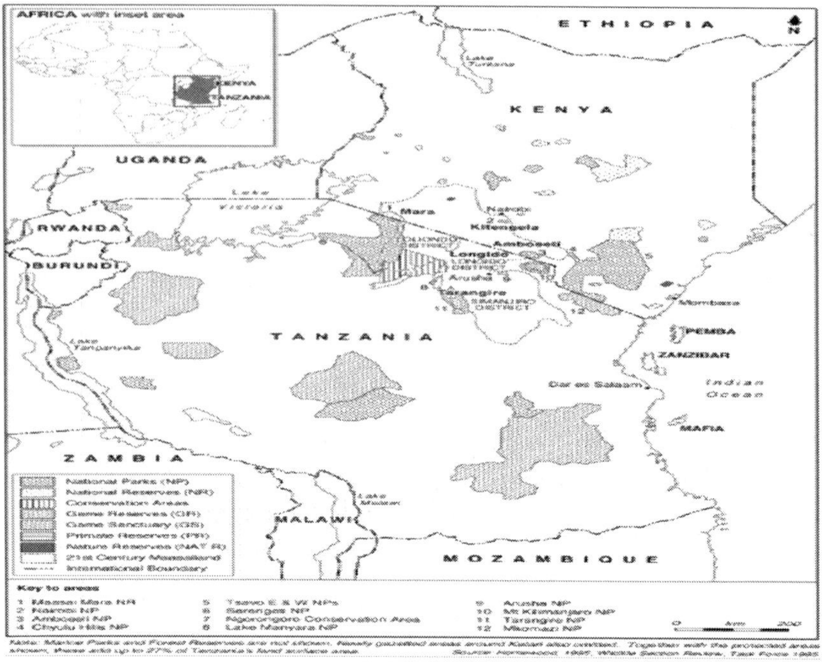

Figure 1: Kenya and Tanzania Maasailand, protected areas and study sites**.**

However, the wildlife populations are in drastic and well-documented decline. Kenya's long-term aerial count datasets show that there has been a 50% to 80% decline in numbers of the vast majority of wildlife species since counts began in the mid-1970s (Ottichilo et al 2001; Homewood et al. 2001; Western et al. 20062009; Ogutu et al. 2011). This decline is as marked within as around protected areas (Western et al. 2006). Though data for other East African regions are less complete, and declines seem less apparent in Tanzania (Homewood et al. 2001), similar trends are emerging (Sinclair et al. 2008). For example, Stoner et al. (2007) collated 1980s to early 2000s aerial census data for large herbivore populations in eight large census zones in Tanzania. Of those that showed significant changes, most declined. Thomson's gazelle, Grant's gazelle, hartebeest, reedbuck, roan antelope, sable antelope, warthog and zebra declined in over 50% of the zones where they were surveyed, though elephant and giraffe generally fared well across the country. Around Tarangire, elephants are increasing, while wildebeest are declining (Sachedina 2008).

These declines have taken place despite heavy investment in wildlife management and conservation by governments and international conservation NGOs (Norton-Griffiths 2007; Scholfield and Brockington 2009). There are a number of possible causes of wildlife decline, which can be categorised using Diamond's 'evil quartet' of factors driving species extinctions. These are habitat loss, introduced species (including disease), overhunting (by human or non-human predators) and secondary extinctions (following the prior extinction of organisms on which the species under consideration depends). In the case of East African wildlife, each of these factors has been shown to apply to some extent for individual species in individual sites. For example, in Serengeti and Ngorongoro, temporary local extinction of wild dog and a drastic decline in lion populations were both caused by an outbreak of canine distemper (Sinclair et al. 2008, page 3). Across Africa, rhinoceros are perennially vulnerable to illegal hunting. Some rhino subspecies are thought to be locally extinct in a number of sites, and others have been hunted to near-complete or complete regional extinction across East Africa (Sinclair et al. 2008), page 39; (Milliken et al.2009). On Laikipia Plateau in Kenya, predator increases associated with private conservancies and their interplay with livestock ranching are thought to have driven down wild herbivore numbers in ways that impact unevenly across numerically dominant as opposed to relatively less numerous species (Georgiadis et al. 2007). More generally, decades of ecological research have shown that forage scarcity, predator pressure, outbreaks of disease and other factors acting whether singly or in concert have all limited the populations of individual East African savannah herbivores (Sinclair et al. 2008). However, both at the global level (Lambin and Meyfroidt 2011; Mace et al. 2005) and more specifically in Kenyan and other East African rangelands (Western et al. 2009), habitat loss is increasingly emerging as the most salient cause of species decline. The marked synchronous decline of regional populations of over 40 large mammal East African savanna species between 1975 and the present day, while exacerbated by the interplay of numerous individual factors, is likely to be driven primarily, though not exclusively, by loss of habitat through land fragmentation and conversion to cultivation. Species decline across the board around Mara, and across Kenya more generally (Western et al. 2009), has been linked to proliferation of large-scale commercial cultivation on the one hand (Homewood et al. 2001), and of smallholdings on the other (Lamprey and Reid

Mara	Maasai Mara NR earns $15 to $20 million pa	Numerous conservancy/ WA/ tourism operations; 64% of hh earn from wildlife, which averages >20% mean annual income		Richest site in livestock; commercial cultivation/ land leasing declining
Kitengela	Nairobi NP	KILA operates leaseback to keep rangeland unfenced; 14% of hh getca 3% mean income this way	Urban effect: Land values > Kshs 100,000/acre Population times 2.5 from 1989 to 1999	Off farm opportunities; urban market for milk (1/3 livestock-related income) and crops
Amboseli	Amboseli NP/NR	Numerous small tourism operations bring little to outer, drier group ranches		Wetlands converted to intensive production; dry land cropping is risky
Longido	None, but ringed by protected areas	Enduimet WMA sets aside majority of land for seven villages: State captures returns, negligible trickledown	Poorer hh rely on (limited) wage opportunities for food and to rebuild herds	Poorest site Farming essential; livestock are few but central
Tarangire	Tarangire and Manyara NPs = 13% area; $4.5 million pa gate fees, $0.5 million hunting revenues	Emboreet CBC; Manyara Ranch Ongoing conflict:Sachedina (2008)	35% of hh get mining-related remittances; 45% respondents have been involved in mining	Mechanised maize cultivation initially tenure strategy, now for commercial profit

CBC community-based conservation, *hh* household, *KILA* Kitengela landowners' association, *NP* national park, *NR* national reserve, *TLU* Tropical livestock unit, *WA* wildlife association (Kenya), *WMA* wildlife management area (Tanzania).

Homewood *et al.*

Homewood *et al. Pastoralism: Research, Policy and Practice* 2012 2:19, doi: 10.1186/2041-7136-2-19

METHODS

The data and comparative analyses outlined here build upon published standardised income and explanatory variables which were collected or derived for each of the study sites during the period 2004 to 2006 by independent researchers[i] and their teams, according to standardised definitions and approaches as agreed during the DGIC-funded Reto-o-Reto project (Serneels et al. 2009). Those data were based upon household surveys, with sampling which used wealth ranking and, in some cases, stratification of households by spatial location. In total, over 1,000 households were surveyed (Table 1) representing a broad cross-section of Maasai in pastoral rangelands? Our interpretation and analysis also draw upon published cluster analyses which identified groups of households with similar patterns of activities in each area and regression analyses which identified significant factors explaining variation in income levels across households (Serneels et al. 2009). Our analysis and interpretation also incorporate insights from published family portrait studies which captured qualitative, household-level pictures of livelihoods and livelihood change (Homewood et al. 2009).

RESULTS AND DISCUSSION

Great variability in livelihood strategies and in wealth was expected (and found) both within and between sites. Briefly, each site had a cluster of livelihood strategies corresponding to households which were primarily dependant on pastoralism (*cf.* households 'staying' as primarily pastoralist; McPeak et al. 2012). Beyond this, each site showed a different array of clusters or livelihood strategies representing different combinations of activities in different relative proportions dictated by local circumstances and opportunities. This great variability is summarised here under livestock, farming, off-farm employment and wildlife-related activities.

Livestock

Pastoral production accounted for well over half of the mean income for the pooled samples in each site (Figure 1). While most (91% to

100%) households had livestock, a significant proportion in each site had too few to support household members fully. Most animals were concentrated in the hands of a few, with the top 10% to 20% households owning half to two-thirds of all livestock across all sites (Table 1).

Reliance on non-livestock income was therefore a necessity for most, especially for the poorest, quite apart from being a potentially positive investment option for the well-off. Nonetheless, in each site, across all different wealth categories and across most livelihood strategies, people were actively purchasing livestock. Poor households continued to seek to rebuild their herds, while better-off households continued to invest in new animals.

Farming

Farming was widely practised. Maasailand is mostly semi-arid, and most households had limited access to agro-ecologically favourable sites. In four out of the five study areas, >50% of households farmed (Kitengela 68%, Amboseli 57%, Longido 67%, Tarangire 88%; see Figure 2). In Mara, only 13% households cultivated.

However, yields were generally poor and contributed little to overall incomes. In Mara, Longido and Kitengela, >50% of the households that farmed harvested nothing. Amboseli yields were variable; again, many households failed to harvest. Crops accounted for just 2% of the mean total income in Mara, 8% in Kitengela, 12% in Tarangire, 14% in Amboseli and 21% in the poorest site, Longido.

Farming contributed to livelihood security on several levels. In addition to the benefits for food security and opportunities for commercial cropping in some sites (Mara, Tarangire), cultivation has been an effective means of staking claim during land privatisation and subdivision e.g. Kenya, Mara; (Thompson et al. 2009), and countering protected area expansion e.g. Tanzania, Tarangire; (Sachedina and Chenevix Trench 2009).

Both direct observation and remotely sensed land cover analyses showed extensive large-scale commercial cereal cultivation in the outer ring of former group ranches around the Maasai Mara National Reserve (MMNR). However, both the proportion of households involved in commercial cultivation and its total acreage dropped significantly between 1998 and 2004 with the completion of land titling across most of the area studied and more recently with the rise of conservancy agreements across the former group ranches immediately adjacent to MMNR (Bedelian in press). Alongside poor rainfall and declining soil fertility, the transaction costs of dealing with multiple smallholders (as opposed to dealing with the group ranch committee for large areas) made large-scale farming in Kenya Maasailand difficult.

Large-scale cereal farming had also spread around Tarangire. Maize cultivation was a lucrative possibility for households able to invest in mechanised farming (Sachedina 2008). However, opportunities for such large-scale farming were not widespread elsewhere. In the poorest site, Longido, with the highest proportion of households farming, former village-owned high-potential lands on the slopes of Mt. Kilimanjaro (east of Longido town) had long since been leased by the state to outside investors (Chenevix Trench et al. 2009).

Off-farm Activities

Half or more households (50% to 85%) earned off-farm income from petty trade, business, wages or salaried income and remittances (Figure 2). Returns from casual unskilled work are a fraction of those for regular jobs as teacher, driver or government official. Potentially large but ephemeral income streams from gemstone mining and brokerage, and land leasing, tended to be seen as secondary in importance to livestock and other economic activities. Off-farm work accounted on average for 8% Mara, 20% Amboseli, 30% Tarangire, 34% Longido and 43% Kitengela income - second only to livestock in most sites, other than Mara (Figure 2). Earnings from off-farm work are particularly salient in two sites: the poorest site Longido (where the poorest households depend entirely on off-farm income or remittances; Figure 3) and peri-urban Kitengela, where there are increased opportunities for such work. This bears out analyses emphasising the need for off-land work and the willingness of pastoral peoples to pursue these activities e.g. Sandford 2006; Boku 2008; Galvin 2009; McPeak et al. 2012).

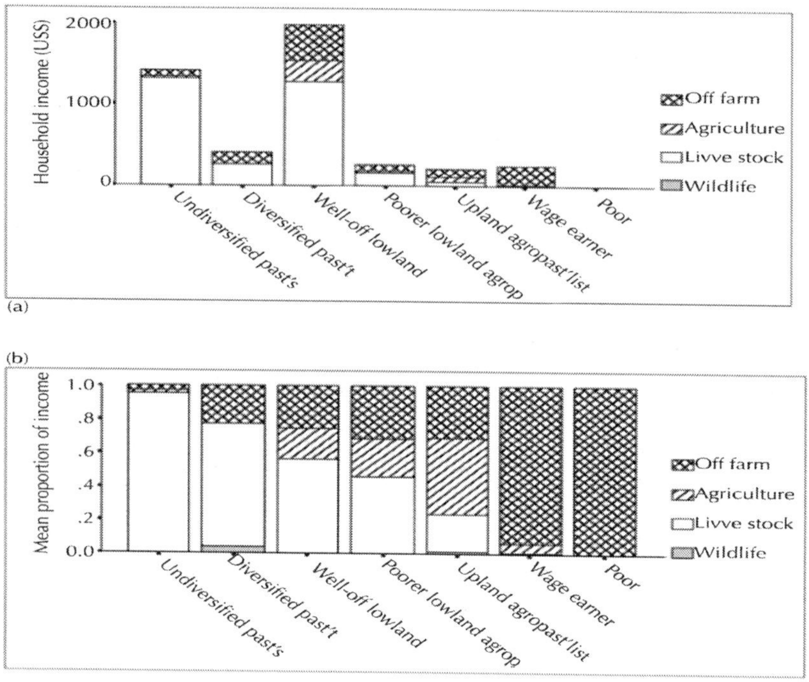

Figure 3: Livelihood strategies in Longido. (a) Different strategies are represented according to annual household income from livestock, agriculture, conservation and off-farm sources. The first bar of the chart represents undiversified pastoralists; the others have significant proportional contributions from other activities. Some households ('poor') had no source of income or production, other than remittances from absent members. (b) Same data represented as proportions of mean total income for each strategy. Note that most have no wildlife-based income, but only the poorest have no livestock.

Wildlife-related Income

There are wildlife-based initiatives in all sites and in most cases, established protected areas too (Table 2). In principle, there are multiple channels whereby local people may derive income from wildlife-related activities. These divide into returns at community level and those which come direct to the household or individual. Possible wildlife-related income streams differ between Kenya and Tanzania, and between different sites within countries.

In Kenya, MMNR gate fees flow to the Narok County Council with a proportion translating into community development. In some sites, a proportion of gate fees earned by other protected areas (national parks, Ngorongoro Conservation Area) are also said to be channelled to community development (but often with little transparency or accountability; see e.g. Sinclair et al. 2008, pages 452 to 454). Returns at the community level are supposed to translate into improvements in health, education, transport and other community facilities though they may be dissipated through poor central, district- or community-level governance, elite capture or outright corruption (Thompson and Homewood 2002; Sachedina 2008). In some cases, tour companies pay contractors directly to carry out village-approved improvements (Sachedina and Chenevix Trench 2009). In both Kenyan and Tanzanian sites, some individuals received bursaries for education or medical treatment either through community board decisions (Thompson and Homewood 2002) or direct from tour operators (Sachedina and Chenevix Trench 2009).

However, with land privatisation around Mara and Kitengela in Kenya, former community-based organisations (e.g. wildlife associations formerly congruent with group ranches around Mara) have largely ceased to function. Instead, they have given way to one-on-one, standardized land lease deals between individual landowners and conservancy enterprises around Mara, and conservation easements around Kitengela. Some landowners have privately owned tourist campsites on their land (Thompson et al. 2009).

In Tanzania, bed-night payments, concession fees and contributions from game viewing or hunting companies could originally be paid directly into village accounts for social development projects, but the new regulations for wildlife management areas (WMAs) mean that these monies are intercepted at the level of central government (Sachedina and Chenevix Trench 2009). Wildlife-related income has in the past been used on occasion at village level to offset cesses (village-level taxes) in Tanzania. This created positive but arguably invisible impacts on household income, in that these community-level effects are not captured by household survey. However, where data were available, it was clear that village-level benefits were easily captured by local elites and were not having the broader impacts on livelihoods that could influence household decision-making in favour of conservation-compatible land use (Sachedina 2008; Nelson et al. 2009).[iii]

[iii]This is not universally the case across Maasailand, and there have been instances of community-based tourism resulting in meaningful revenues that were then well distributed (Nelson and Makko 2003).

Household surveys in all sites focused on income streams coming to individuals within the household. Wildlife-related income was recorded to include activities that would otherwise be categorised as returns from business or salaries but which would not be available as opportunities without the wildlife (Serneels et al. 2009). Some individuals secured employment as labourers or guides e.g. (Sachedina and Chenevix Trench 2009), as community scouts monitoring set-aside areas (e.g. Kitengela; Nkedianye et al. 2009), and as Kenya Wildlife Service rangers (Thompson and Homewood 2002; Thompson et al. 2009). Some individuals derived income from tourist-oriented activities, such as cultural boma performances, sales of beadwork, jewellery and other craftwork. In at least one site, some villagers had received payment for facilitating illegal hunting, though such activities could not be systematically researched through household survey.

To summarise, compared with the universal engagement with livestock and the widespread involvement with farming and off-farm work, only a small proportion of households received wildlife earnings in most sites (3% to 14%). Averaging across those households which do derive income from wildlife, amounts were small in most sites, contributing <5% of the mean annual income (Figure 2).

While overall few Maasai households earn from wildlife, and the sums they make do not compare with main income streams from livestock, crops and off-farm sources, landowning households close to the MMNR saw real benefits from conservation-based enterprises. In contrast to the other sites, two-thirds (64%) of the households in Mara earned some income from wildlife, and wildlife conservation accounted for 21% of the mean annual income for Mara households in the sample (Figures 2 and 4). This is explored in more detail in the next section.

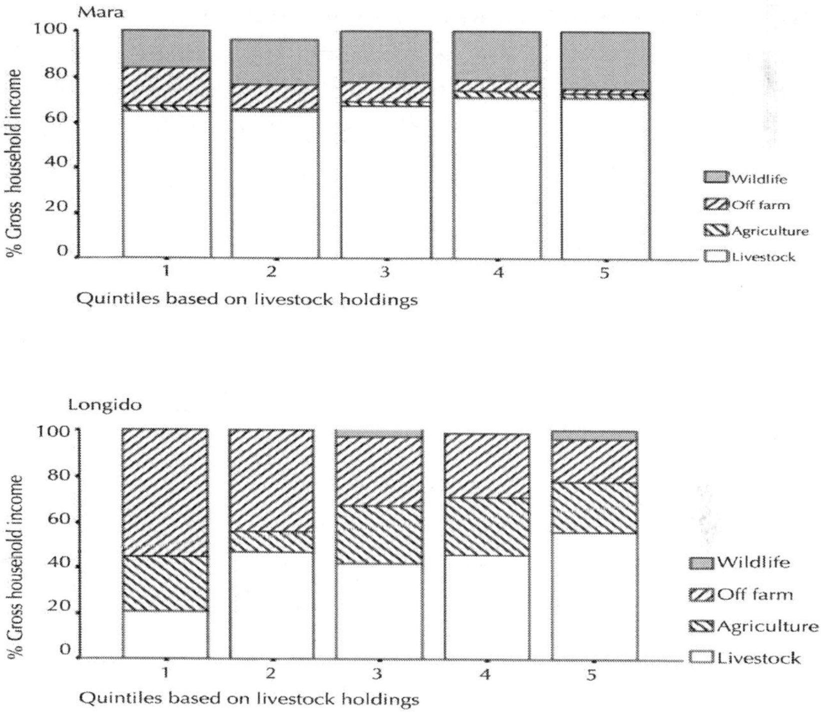

Figure 4: Relative importance of different income sources across different wealth quintiles for households. (a) Mara and (b) Longido households; (1 = poorest/least livestock, 5 = most livestock).

Conservation, Wealth and Poverty in Mara

MMNR is Kenya's highest-earning wildlife tourist destination, taking 15 to 20 million USD annually (Norton-Griffiths 2007). Numerous additional wildlife enterprises have grown up around Mara, with landowners[iv] on the now-adjudicated, subdivided and privately owned former group ranch lands able to capture wildlife returns. There has been a rapid evolution of revenue-sharing institutions around Mara since the 1970s, from the initial 19% county council MMNR gate takings paid to group ranch wildlife associations; through post-group ranch, politically constituted wildlife associations; and most recently, conservancy partnerships between tourism investors and landowners (Thompson et al. 2009; Bedelian in press). The most recent conservancy agreements offer better security of income to landowners, through rent rather than bed-night payments. They require 5- to 15-year covenanting of the designated area, during which current resident households must move out. In covenanted areas, land sales, homestead construction, cultivation and fencing are excluded, as is grazing of livestock under most circumstances (Bedelian in press). Conservancy arrangements may offer a better deal for landowners (Thompson et al. 2009) and better conservation outcomes (Western et al. 2006). They remain to be evaluated in terms of implications for wildlife populations (Western et al. 2009), impacts on households forced to move by conservancy restrictions (particularly those relocating herds, and those who are non-landowners and who therefore receive no compensatory revenues), and impacts on receiving areas (Bedelian in press).

[iv]This paper does not go into the process of privatisation, which dispossessed many vulnerable families (Galaty 1999) but focuses on the impact of conservation business on current, mostly landowning residents.

In Mara, wildlife revenue made up 15% to 30% of the mean household income across the wealth spectrum from the poorest quintiles to the best-off (Figure 4), second in importance only to livestock. However, the top 25% of Mara households by wealth consistently captured 60% to 70% of conservation income. The bottom 25%, by contrast, captured around 5%, rising to 15% if all forms of associated conservation-related employment are included. The middle 50% took around 25% of conservation-related income across the board. Despite significant changes in the volume of tourism returns between 1998 and 2004, there was minimal change in this pattern of distribution across wealth ranks. The poorest 20% of households were consistently more likely to be engaged in cultivation and/or off-farm work and significantly less likely to receive wildlife income than other households (Thompson et al. 2009).

The total volume of tourism returns in Kenya fell significantly between 1998 and 2004, probably as a result of the impact on tourism internationally of the attacks on the World Trade Center in New York in 2001. Tourism collapsed again in 2007 to 2008 following Kenya's post-election violence. In the pre-conservancy era 2004 Mara surveys, mean conservation incomes to households earning from what were then wildlife associations and campsites averaged just 25% of their 1998 value. Within that changing flow of revenue, the relative proportions captured by the wealthiest, middle and poorest Mara households stayed remarkably constant. However, between 1998 and 2004, proportions of households receiving income from wildlife associations fell from 55% to 37%. Overall, the proportion of households reporting income from the wildlife associations and campsites (which were the main wildlife-based initiatives at the time) dropped from 55% to 41%.

Despite inequalities, conservation earnings reached most households in the Mara sample, and returns at household level, while very variable, on average, made a significant contribution. However, comparison of Mara with the four other sites suggests that these benefits are linked to conditions rarely met elsewhere. MMNR is high-earning compared with other tourism destinations. The households sampled are predominantly those of present landowners situated close to the MMNR, who are able to command pay for game viewing or accommodation on their land.[v] By contrast, households elsewhere that are remote from tourist attractions or near lower-earning sites, or do not own land, are less able to compete for conservation jobs or have little access to wildlife income overall. This applies to the outer Amboseli households (Burn Silver 2009); those close to Amboseli National Park reportedly earn significantly more, possibly comparable to those in Mara (Western, personal communication). In the poorest site, Longido, livestock remain the most important livelihood. Additional income streams are dominated by returns from off farm and farming, while wildlife revenues are minimal and poorly distributed across the wealth spectrum (Chenevix Trench et al. 2009; Figures 2 and 3b).

[v]Households which were not able to secure claim to a private plot have been excluded not only from the possibility of such wildlife income, but also from the landscape as a whole (Galaty 1999). By definition, they cannot appear in our sample.

CONCLUSIONS

These findings underline the lasting, central importance of livestock to livelihoods across Maasailand. As for other studies (Boku 2008), livestock holdings represent the single strongest measure or indicator of other dimensions of wealth in all sites.[vi] By contrast, they emphasise both the generally limited contribution of wildlife and conservation income to households and its salience under special circumstances in Mara.

[vi]Tarangire data do not allow for direct comparison.

Diversification is clearly occurring widely across Maasailand but is not necessarily profitable. Non-livestock income presents a potentially positive investment option for the better-off able to access high-quality, well-watered farmland, or relatively high-paying jobs, or to intensify their livestock production. For the poorest families, non-livestock income represents the only means to achieve food security and the only hope of rebuilding the herd. In practice, however, returns to marginal agriculture, irregular unskilled work and petty trade are so low that these households end up drawing down on their assets rather than building them up. Such poverty is distributed across livelihood strategies: there is no clear combination of activities that does best economically in any given set of circumstances (Homewood et al. 2009).

A recent analysis of livelihoods in the northern Kenya rangelands plots the position of households on the two axes of livestock holdings and cash income (McPeak et al. 2012). According to which quadrant they fall in, households are seen as 'left behind' (with below average holdings and cash income), 'staying' (as pastoralists, with above average livestock holdings but below average cash income), 'moving out' (from pastoralism, with below average livestock holdings and above average cash income) or as successfully 'combining' (pastoralism with other activities, above average for both livestock holdings and cash income). The present analysis seeks to go beyond this by looking in more detail at the relative contributions of individual activities to livelihoods, at the ways these combine to produce livelihood strategies, and at the role of wildlife-based revenues within these strategies.

These findings first highlight the lasting importance of livestock to Maasai households. Livestock remain central to subsistence, to pathways out of poverty and to wealth storage/accumulation strategies, alongside people's need or strategic choices to diversify into non-livestock activities. If we were to consider the social importance of livestock in maintaining social relations and not just their economic value, their significance would be all the greater. This absolutely central economic and livelihood importance of livestock is consistent with findings in the northern Kenya/southern Ethiopia pastoral rangelands

(McPeak et al. 2012). The crucial importance of livestock production in these ASALs for both national economy and local livelihoods remains unacknowledged by most government policies in Kenya and more particularly in Tanzania, and also by many conservation interventions (WISP-IUCN 2007; Behnke and Muthami 2011).

Second, cultivation is widely practised but gives very limited returns. Besides adding to food security, it may be a tenure strategy, curbing land-grabbing (Zoomers 2010; Igoe 2007), including what is perceived locally as hostile encroachment of conservation on customary rangelands (Sachedina 2008).

Third, off-farm income is a very significant component of present-day Maasai livelihoods, usually more so than agriculture, but ranges from poorly paid, insecure, often dangerous work (miners, watchmen, sex workers) to secure jobs with wider political/economic prospects (teachers, MPs).

Fourth, these results suggest that communities portrayed by some as the wealthiest land- and stock-owners in East Africa (Norton-Griffiths and Said 2010) have average incomes far below the dollar-per-day international poverty line and often below national rural poverty thresholds too. Given that these average income values are skewed upwards by a small number of well-off households (Homewood et al. 2009), and median incomes are in most cases around half mean values, poverty remains both wide and deep in Maasai rangelands despite potential land values and tourism earnings (Table 3). The mean and median incomes we documented broadly overlap but in some cases exceed the ranges of incomes in the northern Kenya/southern Ethiopia rangelands, mostly similarly below poverty line (Little et al. 2001)

Table 3: Mean value of total annual gross income in US dollars per household per year (±standard deviation)

Total gross income	Kitengela	Amboseli	Mara	Longido	Tarangire
					(N=26)
$/household/year	2,511 ± 2,497	1,583 ± 1,655	2,625±2,892	733±1,518	2,317±2,150
$/AE/year	453±445	178±186	474±563	74±107	403±380

$/AE/day	1.24	0.49	1.29	0.20	1.10
$/person/ daya	1.78	0.48	0.84	0.16	0.85

AE adult equivalent; [a]Kenya's official poverty line is Kshs 1,239/month (CBS 2003; equivalent to $0.53/person/day).

Homewood *et al.*

Homewood *et al. Pastoralism: Research, Policy and Practice* 2012 2:19, doi: 10.1186/2041-7136-2-19

Last but not least, wildlife generally performs poorly for livelihoods. With the exception of Mara, wildlife brings little or nothing to the vast majority of Maasai. If wildlife does not become locally valuable, it may continue to decline (Norton-Griffiths 2007; Norton-Griffiths and Said 2010; Ogutu et al. 2011).

Why does such a potentially profitable enterprise bring so little local benefit? These communities have historically captured little of total tourism earnings, with approximately 95% accruing to tour operators, service industry workers, and the state (Norton-Griffiths 2007; Norton-Griffiths ET al.2008; Norton-Griffiths and Said 2010). The small amounts captured are then poorly distributed (Homewood and Thompson 2010). Proportions captured by local residents are even lower in Tanzania (Sachedina 2008) as they trickle through official channels, top-sliced at each stage from central and district governments through to the communal level via WMAs.

In part, this reflects chronic problems of governance and accountability at local, district and national levels e.g. (Walpole and Leader-Williams 2001; Sinclair et al. 2008; Homewood 2009; Bedelian in press). In Tanzania, wildlife enterprises earned tens of thousands of dollars annually for one village on the edge of Tarangire, and yet these revenues were easily dominated by local elites (Sachedina 2008). Ololosokwan village, situated east of Serengeti, was able, briefly, to earn around $50,000/year from photographic safari operator use of its lands. However, the central government feared that this arrangement would conflict with a hunting block whose revenues they control. In November 2007, a surprise ministerial declaration criminalised local-level deals for wildlife-related enterprises, capturing all such returns for the state, with no requirement for a set proportion to be returned to the community (TNRF 2007). Such unequal contests between state and local communities for control of conservation enterprises and their returns have become a common occurrence in Tanzania (Nelson 2004; Nelson 2007; Baldus 2009). They are made the more unequal by the involvement of global investors (Igoe 2007). In their comparative study of community-based conservation, Nelson and Agrawal (2008) observe that the hunting industry in Tanzania is eminently corruptible, providing easily diverted revenues within a generally impoverished national economy.

Another reason is the historical experience of many Maa-speaking pastoralists. Conservation, for them, is associated with large-scale eviction and exclusion (from Serengeti, Ngorongoro, Tarangire, Amboseli, Mkomazi; Figure 1) with fines and harassment, compromises and deals that were not honoured and outreach programmes providing few tangible benefits (Brockington 2002). Their experience of new revenue-sharing initiatives is rarely positive, and where it could be (as in some emerging conservancy agreements around Mara) distributional and knock-on effects of displacement are only just beginning to register. The livelihood choices that Maasai face now are shaped by decades of such experiences, perceptions and stories, as well as by complex communal politics, making it hard to build trust and cooperation (West et al. 2006; Adams and Hutton 2007).

A third reason has resulted from the efforts of the conservation lobby itself. The growing financial success of the African Wildlife Foundation went hand in hand with its growing out of touch with dilemmas in Maasai villages (Sachedina 2008, Sachedina 2010). This meant that it was poorly equipped to engage effectively with the fierce local politics that surround conservation initiatives in this region. Engaging with communities with such a record is an extremely difficult task.

Finally, these conservation enterprises have been given tremendous publicity, portrayed by their advocates as win-wins, good for wildlife, good for people, good for the economy, participatory, empowering and liberating (Igoe et al. 2010). Yet, there are few robust and in-depth evaluations of these initiatives, and where they have been examined in detail, these schemes rarely produce the benefits they claim. Once returns are divided up among the total population concerned, apparently high aggregate sums allocated to communities soon become insignificant as one-off payments. Moreover, distribution may often be highly inequitable. As this paper underlines, revenues from wildlife rarely begin to compensate for loss of mobility, access to and control

AUTHORS' CONTRIBUTIONS

This paper represents a synthesis and discussion built on findings from individual studies whose principal researchers are listed in the acknowledgements below, with full details of all involved given in Homewood, Kristjanson and Chenevix Trench 2009. KH drew out the main conceptual findings from the underlying studies and drafted the present paper. PCT managed data, comparative analyses and improved the paper. KH, PCT and DB all contributed to interpretation and intellectual content and worked together to produce the final draft. All authors read and approved the final manuscript.

ACKNOWLEDGEMENTS

This paper draws on material and insights from individual studies by Shauna Burn Silver, Michael Thompson, David Nkedianye, Hassan Sachedina, Dickson Ole Kaelo, Patti Kristjanson, Maren Radeny, Fred Nelson, Stephen Kiruswa, Susanne Serneels, Mario Herrero and others, published in detail in Homewood, Kristjanson and Chenevix Trench 2009. We are grateful to all who collaborated in that work, to the many Maasai households who discussed their livelihoods with the researchers, to the Kenya and Tanzania governments who gave permission for the original fieldwork, and to the Belgian government DGIC fund who made possible the original collaborative work in drawing together these individual studies. A brief preliminary version of this paper was presented at the 2009 Zoological Society of London symposium on Biodiversity conservation and Poverty reduction, and published as *Pastoralism and conservation: who benefits?* In Roe et al. (in press).

REFERENCES

1. Adams, W, and J Hutton. 2007. People, parks and poverty: political ecology and biodiversity conservation. Conservation and Society 5 (2): 147–183.
2. AWF. 2005. Kilimanjaro heartlands: a summary of AWF's engagement and related GIS data. African Wildlife Foundation Presentation given at ILRI Reto-e-Reto conference.January 2005

3. Baldus, R. 2009. A practical summary of experiences after three decades of community-based wildlife conservation in Africa "What are the lessons learnt?" Budapest: FAO and CIC.

4. Bedelian, C. Conservancy land leases in the Mara, Kenya. Nairobi: LDPI working paper.in press

5. Behnke, R, and D Muthami. 2011. The contribution of livestock to the Kenyan economy. IGAD LPI Working Paper No. 03–11. Intergovernmental authority for development in Eastern Africa. Ababa: IGAD LPI Communications Office.

6. Bishop, E. 2007. Schooling and pastoralists' livelihoods: a Tanzanian case study. London: University College London. Unpublished PhD thesis

7. Boku, T. 2008. Pastoralism under stress: resources, institutions and poverty among the Borana Oromo in Southern Ethiopia. Norwegian University of Life Sciences.PhD thesis, Ås

8. Boone, RB, KA Galvin, SB BurnSilver, PK Thornton, DS Ojima, and JR Jawson. 2011. Using coupled simulation models to link pastoral decision making and ecosystem services. Ecology and Society 16 (2): 6.

9. Brockington, D. 2002. Fortress conservation: the preservation of the Mkomazi Game Reserve. Oxford: James Currey Publishers.

10. Brockington, D. 2006. The politics and ethnography of environmentalisms in Tanzania. African Affairs 105 (418): 97–116.

11. Brockington, D, and K Scholfield. 2010. Conservation NGOs and the conservationist mode of production in sub-Saharan Africa. Antipode 42 (3): 551–575

12. Bryceson, DF, and V Jamal. 1997. Farewell to farms: de-agrarianisation and employment in Africa. Aldershot: Ashgate.

13. BurnSilver, S. 2009. Pathways of continuity and change: Maasai livelihoods in Amboseli, Kajiado District, Kenya. In Staying Maasai? Livelihoods, Conservation and Development in East African Rangelands, ed. Homewood K, Kristjanson P, Chenevix Trench P, 161–208. New York: Springer.

14. Castillo, AR. 2004. Sustainable inequalities? Community-based conservation in an 'inequitable society'. The case of Il Ngwesi Group Ranch, Kenya. Master's Thesis. Oxford: University of Oxford.

15. CBS. 2003. Annual facts and figures. Central Bureau of Statistics, Government of Kenya, Nairobi (now Kenyan National Central Bureau of Statistics KNBS).
16. Chenevix Trench, P, S Kiruswa, F Nelson, and K Homewood. 2009. Still people of cattle? Livelihoods, diversification, and community conservation in Longido District. In Staying Maasai? Livelihoods, Conservation and Development in East African Rangelands, ed. Homewood K, Kristjanson P, Chenevix Trench P, 217–256. New York: Springer.
17. Davies, J, and M Nori. 2008. Managing and mitigating climate change through pastoralism. IUCN Commission on Environmental, Economic and Social Policy: Policy Matters.http://cmsdata.iucn.org/downloads/policy_matters_16.pdf
18. Dietz, T, AA Nunow, AW Roba, and F Zaal. 2001. Pastoral commercialisation: on caloric terms of trade and related issues. In African Pastoralism: Conflict, Institutions and Government, ed. Salih MA, Dietz T, Ahmed AGM, 194–234. London/ SterlingVirginia: Pluto Press/OSSREA.
19. Galaty, J. 1999. Grounding pastoralists: law, politics and dispossession in East Africa. Nomadic Peoples 3 (2): 56–73.
20. Galvin, KA. 2009. Transitions: pastoralists living with change. Annual Review of Anthropology 38 (1): 185–198.
21. Georgiadis, N, F Ihwagi, JGN Olwero, and S Romanach. 2007. Savanna herbivore dynamics in a livestock-dominated landscape. II. Ecological, conservation, and management implications of predator restoration. Biological Conservation 137: 473–483.
22. Homewood, K. 2008. Ecology of African pastoralist societies. Oxford: James Currey and Ohio UP.
23. Homewood, K. 2009. In Staying Maasai? Livelihoods, Conservation and Development in East African Rangelands, ed. Homewood K, Kristjanson P, Chenevix Trench P, 335–368. New York: Springer.
24. Homewood, K, and M Thompson. 2010. Social and economic challenges to conservation in East African rangelands: land use, livelihoods and wildlife change in Maasailand. In Wild Rangelands: Conserving Wildlife While Maintaining Livestock in Semi-arid Ecosystems, ed. Deutsch J, Toit J, Kock R, 340–366. Oxford: Wiley Blackwell.

25. Homewood, K, E Lambin, E Coast, a Kariuki, I Kikula, J Kivelia, M Said, S Serneels, and M Thompson. 2001. Long-term changes in Serengeti-Mara wildebeest and land cover: pastoralism, population or policies? Proceedings of the National Academy of Sciences 98 (22): 12544–12549.

26. Homewood, K, M Thompson, PC Trench, S Kiruswa, and E Coast. 2005. Community- and state-based natural resource management and local livelihoods in Maasailand. Gestione della risorse naturali su base communitaria e statale. Ambiente e sviluppo sostenibile in Africa australe. Special issue on community-based natural resource management. Afriche e Orienti 2005 (2): 84–101.

27. Homewood, K, P Trench, S Randall, L Lynen, and E Bishop. 2006. Impacts of a veterinary intervention in Maasai Cattle: livestock health, socio-economic correlates and ecological implications. The infection-and-treatment vaccine against East Coast fever. Agricultural Systems 89: 248–71.

28. 2009. Staying Maasai? Livelihoods, Conservation and Development in East African Rangelands,Homewood K, Kristjansen P, Chenevix Trench P New York: Springer.

29. Homewood, K, D Brockington, and S Sullivan. 2010. Alternative view of Serengeti road. Nature 467: 788–789

30. Hutton, J, W Adams, and J Murombedzi. 2005. Back to the Barriers? Changing narratives in biodiversity conservation. Forum for Development Studies 2: 341–370.

31. Igoe, J. 2007. Human rights, conservation and the privatization of sovereignty in Africa – a discussion of recent changes in Tanzania. Policy Matters 15: 241–254.

32. Igoe, J, and B Croucher. 2007. Conservation, commerce, and communities: the story of community-based wildlife management areas in Tanzania's northern tourist circuit. Conservation and Society 5 (4): 534–561.

33. Igoe, J, K Neves, and D Brockington. 2010. A spectacular eco-tour around the historic bloc: theorizing the current convergence of conservation and capitalism. Antipode 42 (3): 486–512.

34. Iliya, M, and K Swindell. 1997. Winners and losers: household fortunes in the urban peripheries of Northern Nigeria. In Farewell

to Farms. De-agrarianisation and Employment in Africa, ed. Bryceson D, Jamal V, 85–100. Aldershot: Ashgate.

35. Lambin, EF, and P Meyfroidt. 2011. Global land use change, economic globalization, and the looming land scarcity. Proceedings of the National Academy of Sciences (PNAS) early edition 108: 3465–3472.
36. Lamprey, RH, and RS Reid. 2004. Expansion of human settlement in Kenya's Maasai Mara: what future for pastoralism and wildlife? Journal of Biogeography 31 (6): 997–1032.
37. Lewa Wildlife Conservancy. 2010. Lewa wildlife conservancy: a catalyst for conservation. http://www.lewa.org Accessed 14 March 2012
38. Little, P. 2003. Somalia: economy without a State. Oxford: James Currey.
39. Little, PD, K Smith, BA Cellarius, D Layne Coppock, and CB Barrett. 2001. Avoiding disaster: diversification and risk management among East African herders. Development and Change 32 (3): 401–433.
40. Mace, R. 1993. Transitions between cultivation and pastoralism in sub-Saharan Africa. Current Anthropology 34 (4): 363–381.
41. Mace, G, H Masundire, and J Baillie. 2005. Biodiversity. In Ecosystems and Human Well-being: Current State and Trends: Findings of the Condition and Trends Working Group (millennium ecosystem assessment series), ed. Ceballos G, Lavorel S, Orians G, Pacala S, 74–115. Washington, D.C.: Island Press.
42. 2006. Pastoral livestock marketing in Eastern Africa: research and policy challenges,McPeak J, Little P London: Intermediate Technology Publications.
43. McPeak, J, P Little, and C Doss. 2012. Risk and social change in an African rural economy: livelihoods in pastoral communities. London: Routledge.
44. Milliken, T, R Emslie, and B Talukdar. 2009. African and Asian rhinoceroses – status, conservation and trade. CoP15 Doc. 45.1 Annex. http://www.cites.org/common/cop/15/doc/E15-45-01A.pdf . Accessed 14 March 2012
45. Ministry of Tourism. 2010. Facts and figures. http://www.tourism.go.ke/ministry.nsf/pages/facts_figures Accessed 14 March 2012

46. Ministry of Tourism and Wildlife, Republic of Kenya. 2006. Statistical analysis of tourism trends. Central planning Unit. http://www.tourism.go.ke/ministry.nsf/doc/Tourism_Trends_OCT2006_Revised.pdf Accessed November 2008

47. MoL. 2007a. Draft national land policy. National land policy secretariat. 61. Nairobi: Ministry of Lands.

48. MoL. 2007b. National land policy: the formulation process. National land policy secretariat. Nairobi: Ministry of Lands and Daily Nation.

49. MoLF. 2006. Draft national livestock policy. Nairobi: Ministry of Livestock and Fisheries Development.

50. MTW. 2007. Draft wildlife bill 2007. Nairobi: Ministry of Tourism and Wildlife.

51. Nelson, F. 2004. The evolution and impacts of community-based ecotourism in northern Tanzania. Drylands issue paper no. 131. London: IIED.

52. Nelson, F. 2007. Emergent or illusory? Community management in Tanzania. IIED issues paper no. 146: 1–32.

53. Nelson, F, and A Agrawal. 2008. Patronage or participation? Community-based natural resource management reform in sub Saharan Africa. Development and Change 39 (4): 557–585.

54. Nelson, F, and S Makko. 2003. Communities, conservation and conflicts in the Tanzanian Serengeti. Savannah Georgia: Proceedings of the third annual community-based conservation network seminar: turning natural resources into assets.

55. Nelson, F, B Gardiner, J Igoe, and A Williams. 2009. Community-based conservation and Maasai livelihoods in Tanzania. In Staying Maasai? Livelihoods, Conservation and Development in East African Rangelands, ed. Homewood K, Kristjansen P, Chenevix Trench P, 299–333. New York: Springer.

56. Nkedianye, D, M Radeny, P Kristjanson, and M Herrero. 2009. Assessing returns to land and changing livelihood strategies in Kitengela. In Staying Maasai? Livelihoods, Conservation and Development in East African Rangelands, ed. Homewood K, Kristjansen P, Chenevix Trench P, 115–149. New York: Springer.

57. Norton-Griffiths, M. 2007. How many wildebeest do you need? World Economics 8 (2): 41–64.

58. Norton-Griffiths, M, and M Said. 2010. The future for wildlife on Kenya's rangelands: an economic perspective. In Wild Rangelands: Conserving Wildlife While Maintaining Livestock in Semi-arid Ecosystems, ed. Deutsch J, Toit J, Kock R, 367–392. Chichester: Wiley Blackwell.

59. Norton-Griffiths, M, M Said, S Serneels, D Kaelo, M Coughenour, R Lamprey, M Thompson, and R Reid. 2008. Land-use economics in the Mara area of the Serengeti ecosystem. In Serengeti III: Human Impacts on Ecosystem Dynamics, ed. Sinclair ARE, Packer C, Mduma S, Fryxell J, 379–426. Chicago: Chicago University Press.

60. Ogutu, J, N Owen-Smith, H-P Piepho, and M Said. 2011. Continuing wildlife population declines and range contraction in the Mara region of Kenya during 1977–2009. J Zoology. 285: 99–109.

61. Olajammi, S. 2006. Contested lands: land disputes in semi-arid parts of Northern Tanzania: case studies of the Loliondo and Sale divisions in the Ngorongoro District. Doctoral dissertation: University of Helsinki.

62. Ottichilo, W, J de Leeuw, A Skidmore, H Prins, and M Said. 2001. Population trends of large non - migratory wild herbivores and livestock in the Masai Mara ecosystem, Kenya, between 1977 and 1997. African Journal of Ecology 38: 202–216.

63. Oxfam. 2002. Milking the CAP: how Europe's dairy regime is devastating livelihoods in the developing world. Washington: Oxfam International.

64. Oxfam. 2006. Delivering the agenda: addressing chronic under-development in Kenya's arid lands. Oxfam International Briefing Paper, May 2006

65. Pearce, D, and D Moran. 1994. The economic value of biodiversity. London: Earthscan.

66. Roe, D, M Walpole, and J Elliott. Linking Biodiversity Conservation and Poverty Reduction. Wiley-Blackwell. Conservation Science and Practice series. in press

67. Sachedina, H. 2008. Wildlife is our oil: conservation, livelihoods and NGOs in the Tarangire ecosystem. University of Oxford: Unpublished PhD thesis.

68. Sachedina, H. 2010. Disconnected nature: the scaling up of African Wildlife Foundation and its impacts on biodiversity conservation and local livelihoods. Antipode 42 (3): 603–623.

69. Sachedina, H, and P Chenevix Trench. 2009. Staying Maasai? Livelihoods, Conservation and Development in East African Rangelands. In Cattle and crops, tourism and Tanzanaite: poverty, land use change and conservation in Simanjiro District, Tanzania, ed. Homewood K, Kristjanson P, Chenevix Trench P, 263–298. New York: Springer.

70. Sandford, S. 2006. Foreword in: pastoral livestock marketing in Eastern Africa: research and policy challenges, eds. J McPeak and P Little. Warwickshire: ITDG Publishing.

71. Scholfield, K, and D Brockington. 2009. Non-governmental organisations and African wildlife conservation: a preliminary analysis. BWPI working paper 80. Newcastle and Manchester Universities.

72. Sellen, DW. 2003. Nutritional consequences of wealth differentials in east African pastoralists: the case of the Datoga of northern Tanzania. Human Ecology 31 (4): 529–570.

73. Serneels, S, M Herrero, S BurnSilver, P Chenevix Trench, K Cochrane, K Homewood, P Kristjanson, F Nelson, M Radeny, M Thompson, and M Said. 2009. Methods in the analysis of Maasai livelihoods. In Staying Maasai? Livelihoods, Conservation and Development in East African Rangelands, ed. Homewood K, Kristjanson P, Chenevix Trench P, 43–68. New York: Springer.

74. Sikana, P, C Kerven, and R Behnke. 1993. from subsistence to specialised commodity production: commercialisation and pastoral dairying in Africa, pastoral development network 34: 1–46. London: Overseas Development Institute.

75. Sinclair, ARE, C Packer, S Mduma, and J Fryxell. 2008. Serengeti III: human impacts on ecosystem dynamics. Chicago: Chicago University Press.

76. Stoner, C, T Caro, S Mduma, C Mlingwa, G Sabuni, and M Borner. 2007. Assessment of effectiveness of protection strategies in Tanzania based on a decade of survey data for large herbivores. Conservation Biology 21: 635–646. PubMed Abstract |

77. Swift, J. 1986. The economics of production and exchange. In Pastoralists of the west African Savanna, ed. Adamu M, Kirk-Greene A, 175–190. Manchester: Manchester University Press.

78. Tenga, R, A Mattee, N Mdoe, R Mnenwa, S Mvungi, and M Walsh. 2008. A study on the options for pastoralists to secure their livelihoods in Tanzania. CORDS, PWC, IIED, MMM Ngaramtoni, TNRF, UCRT. http://www.tnrf.org/node/7487

79. The Economist. 1993. The fat of the land: made in Europe, dumped in Africa.

80. The Tanzania National Website. 2009. The economic survey 2009. http://www.tanzania.go.tz/economicsurveyf.html, page 139 . Accessed 14 March 2012

81. Thompson, DM, and K Homewood. 2002. Elites, entrepreneurs and exclusion in Maasailand. Human Ecology 30: 107–138.

82. Thompson, M, S Serneels, and EF Lambin. 2002. Land-use strategies in the Mara ecosystem (Kenya): a spatial analysis linking socio-economic data with landscape variables. In Linking People, Place and Policy: a GIScience Approach, ed. Walsh SJ, Crews-Meyer KA, Boston: Kluwer Academic

83. Thompson, M, S Serneels, D Ole Kaelo, and P Chenevix Trench. 2009. Maasai Mara – land privatization and wildlife decline: can conservation pay its way? In Staying Maasai? Livelihoods, Conservation and Development in East African Rangelands, ed. Homewood K, Kristjanson P, Chenevix Trench P, 77–114.

84. Thornton, P, S BurnSilver, R Boone, and K Galvin. 2006. Modeling the impacts of group ranch subdivision on agro-pastoral households in Kajiado, Kenya. Agricultural Systems 87: 331–356.

85. TNRF. 2007. New regulations signed for all non-consumptive wildlife use in game reserves and village lands. Tanzania Natural Resources Forum. http://www.tnrf.org/node/6529 . Accessed 14 March 2012

86. UNDP/UNEP/IIED/IUCN/WRI. 2005. Environment for the MDGs. A message to the World Summit. Sustaining the environment to fight poverty and achieve the MDGs: the economic case and priorities for action. New York: UNDP. http://www.povertyenvironment.net/pep

87. URT. 1997. Livestock and agriculture policy. Section 3: soil conservation and land use planning. pp 55–68. Section 4: range management pp127-131. Dar es Salaam, Tanzania: Policy statements, United Republic of Tanzania.

88. URT. 2002. The United Republic Of Tanzania Ministry Of Natural Resources and Tourism. Tourism master plan: strategy and actions. http://www.tzonline.org/pdf/tourismmasterplan.pdf

89. URT. 2005. Mkukuta: Tanzania's national strategy for growth and reduction of poverty. Vice-President's Office: United Republic of Tanzania.

90. Walpole, MJ, and N Leader-Williams. 2001. Masai Mara tourism reveals partnership benefits. Nature 413: 771

91. West, P, J Igoe, and D Brockington. 2006. Parks and peoples: the social impact of protected areas. Annual Review of Anthropology 35: 14.1–14.27.

92. Western, D, S Russell, and K Mutu. 2006. The status of wildlife in Kenya's protected and non-protected areas. Paper commissioned by Kenya's wildlife policy and legislation review. Nairobi: African Conservation Centre

93. Western, D, S Russell, and I Cuthill. 2009. The status of wildlife in protected areas compared to non-protected areas of Kenya. PloS One 4 (7): e6140

94. WISP-IUCN.2007.Global review of the economics of pastoralism. http://data.iucn.org/wisp/documents_english/global_review_ofthe_economicsof_pastoralism_en.pdf

95. Zaal, F, and T Dietz. 1999. of markets, meat, maize and milk: pastoral commoditisation in Kenya. In The poor are not us, ed. Anderson D, Broch-Due V, 163–194. Oxford: James Currey.

96. Zaal, F, M Ole Siloma, R Andiema, and a Kotomei. 2006. The geography of integration: cross-border livestock trade in East Africa. In Pastoral Livestock Marketing in Eastern Africa. Research and Policy Challenges, ed. McPeak J, Little P, 145–168. London: Intermediate Technology Publications.

97. Zoomers, A. 2010. Globalisation and the formalisation of space: seven processes driving the current global land grab. Journal of Peasant Studies 37 (2): 429–447

Chapter 7

Users or Producers of Ecosystem Services? A Scenario Exercise for Integrating Conservation and Reindeer Herding in Northeast Finland

Hannu I Heikkinen[1], Simo Sarkki[2], and Mark Nuttall[3]

[1]Faculty of Humanities, University of Oulu, PO Box 1000, 90014, Oulu, Finland

[2]Thule Institute, University of Oulu, PO Box 7300, 90014, Oulu, Finland

[3]Department of Anthropology, University of Alberta 13–15 HM Tory Building, Edmonton, Alberta, T6H 2H4, Canada

ABSTRACT

Nature conservation is often seen as incompatible with pastoralism: conserved predators can cause significant harm to livestock, or livestock may conquer space from natural species. Currently, nature conservation and related policies are increasingly framed through the notion of ecosystem services, which form basis for human well-being. What are the implications of this framing in the relationships between pastoralism and conservation? To answer this question, we draw from our previous research to build four scenarios on the relationships between reindeer herding and nature conservation in northeast Finland. In the two ecosystem service, 'user' scenarios reindeer herders are seen to degrade nature, and thus should be controlled. In the first scenarios, reindeer herding is still compensated for its losses to predation, but in the second scenario, 'reindeer farming' and free grazing practises are eroded for achieving conservation objectives. If herding is viewed as a producer of ecosystem services, the herders' incentivised function would be to sustain predator populations and resemble natural species (wild forest reindeer). In our two 'producer' scenarios, new professions as ecosystem stewards or as ecotourism entrepreneurs emerge and herding changes from meat production priority to supporting the production of other ecosystem services. Novel solutions to conservation-pastoralism dilemmas might emerge if pastoral communities are seen as potential producers of ecosystem services. However, 'producer-based' conservation practises also include various uncertainties and might have negative impacts on pastoral communities' well-being, especially on the continuity of social and cultural traditions.

BACKGROUND

Nature conservation and pastoralism are often considered incompatible and contradictory (e.g. Thompson 1993; Newmark et al. 1994; Buller 2008; Skogen et al. 2008). In many cases, contradiction tends to arise from the restrictions that conservation, and the protected areas are felt to impose on pastoralism and the livelihoods of pastoralists or from the negative effects that predator conservation has on livestock. Often viewed as ecologically and symbolically important, predator animals are becoming the common objectives of nature conservation policies. In northern Finland, there is rising local anxiety that predators such as wolves, brown bears, lynx and wolverines are posing a threat to reindeer herding (Bisi et al. 2007; Sandström et al. 2009; Heikkinen et al. 2011; Hiedanpää and Bromley 2011). In addition, discussions that herding should be restricted in conservation areas are also provoking considerable debate about the rights of reindeer herders and the viability of local livelihoods in the future (Heikkinen et al. 2010; Sarkki2011; Sarkki et al. In press). More widely in Europe where successful biodiversity conservation has long been considered to be inextricably linked to the existence of protected areas, biodiversity is declining despite widespread protected areas coverage — it is argued that the failure of protected areas management is due largely to its neglect of local people's perspectives and participation and the absence of any benefits to the local communities (Hirschnitz-Garbers and Stoll-Kleemann 2011).

Nature conservation agencies increasingly frame discussions of nature conservation not merely in terms of species or area conservation but with the idea of ecosystem services (Daily 1997; Costanza et al. 1997; Millennium Ecosystem Assessment 2003). Ecosystem services encompass benefits of nature to people, and they are often classified in four categories: provisioning, cultural, regulating and supporting services. The key point is that nature conservation is no longer justified just because of nature's intrinsic value but because of the benefits nature provides for humans. Indeed, the ecosystem service approach is increasingly applied to environmental policies of, for example, the European Union and by many nature conservation organisations, initiatives and non-government organizations (NGOs). The rise of the ecosystem services approach is very much a part of both the rethinking of new conservation measures and new ways of conceptualising human-environment relations by the International Conservation and Policy Communities (see European Union (EU) Nature & Biodiversity Partnerships 2009; WWF (2010); Convention on Biological Diversity (CBD) (2010); Countdown 2010). It is thus an increasingly interesting question as to how this new framing of human-environment relations influences and impacts pastoralism-nature conservation relationships. Will the ecosystem service approach make pastoralism more compatible with conservation efforts than previous attempts? How does the ecosystem service approach affect the relationships between as being a prognosis for the future, but here we embrace a definition that scenarios are more akin to consistent and plausible narratives and storylines about how the future might unfold, as well as the existing explorations of critical uncertainties (Raskin et al. 2005). Scenarios are significant for decision makers and stakeholders concerned with formulating objectives for planning and coping with uncertain futures (Kok et al. 2007). We use scenarios as a method to stimulate critical discussion about how the future might unfold regarding reindeer herding and nature conservation in the context of ecosystem services. As such, we contribute to current discussion about how social sciences such as anthropology can engage in multidisciplinary dialogue about scenarios and storylines for a range of possible futures (e.g., Nuttall 2012). Furthermore, beyond our concern with reindeer herding, and as conservation is increasingly framed with the idea of ecosystem services, we suggest that our scenarios can provide fertile reference points for other cases. In this way, we address in part the call by the Millennium Ecosystem Assessment (MA) for advances in models that link social and ecological processes and scenarios for cultural and supporting ecosystem services (Millennium Ecosystem Assessment2005).

We chose two key uncertainties for creating and distinguishing the four scenarios, both relating to policy choices and their relationship to reindeer herding. The first uncertainty is whether reindeer herding is viewed as a user or producer of ecosystem services, especially by decision makers (as opposed to the view of herders themselves that reindeer herding is a livelihood, as well as being a basis for local identities and cultural forms in many cases). If regarded as a 'user', there is a tendency to regulate herding more as it is perceived that reindeer herding is actually contributing to deteriorating ecosystem services and reducing biodiversity. If it is seen as a 'producer' of ecosystem services, however, reindeer herding is more likely to be supported and valued for its contribution to maintaining and nurturing biodiversity and even lauded for enhancing it. The second uncertainty relates to whether policies either aim to transform free grazing reindeer herding, for example, by top-down policy regulation, or to support existing herding practises, for example, by providing forms of compensation and subsidies.

THE ECOSYSTEM SERVICE APPROACH

Policy Responses to the Challenge of Declining Ecosystem Services

The Millennium Ecosystem Assessment (Millennium Ecosystem Assessment 2005) stated that the loss of biodiversity and ecosystem services is one of the main challenges for human well-being. In addition, the Global Biodiversity Outlook (GBO) 2 set by the Convention on Biological Diversity noted that the coverage of protected areas is increasing, but that almost all other biodiversity indicators are showing negative trends. Habitats and ecosystems are deteriorating, changes in species distribution and abundance are worrying, the status of many threatened species is weakening, ecosystems are becoming more fragmented leading to lack of connectivity, and invasive alien species are taking over space from original species (GBO 2 2006). Hence, it is obvious that current conservation measures are insufficient (Chan et al. 2007). Given this, how are policy communities responding to the challenges posed by declining ecosystem services?

One option to halt the loss of biodiversity and ecosystem services is the creation of strong top-down policies and additional conservation measures as considered for example in Global Environmental Outlook 4's (United Nations' Environmental Program 2007) 'Policy First' scenario. This scenario considers nature-based livelihoods as users of ecosystem services whose actions should be restricted, regulated and controlled with top-down policies to maintain biodiversity and ecosystem services.

Another option to enhance ecosystem services would be to plan and implement various market-based governance instruments in order to develop incentives to maintain and also create ecosystem services. This approach, promoted for example with payments for ecosystem service (PES) schemes, assumes that by recognising and bringing ecosystem services under market regulation, resource use would become more sustainable. Here people are not only to be compensated for losses arising from nature conservation, but also incentivised for producing ecosystem services. By this way, the economic motivation is enhanced for advancing conservation willingness and thus balancing the inequalities in costs and benefits of conservation (Bruner et al.2008; Jack et al. 2008; Sommerville et al. 2009).

Thirdly, the MA's (MA 2005) TechnoGarden scenario is based on the idea that ecosystem services can and should be generated and actively created by people, for example, by various kinds of environmental engineering approaches. Resource users would be incentivised or regulated in order to create transformations in human-environment interactions towards supporting and producing ecosystem services.

The Ecosystem Service Approach as an Analytical Tool

The ecosystem services approach has its origins in environmental economics (Farber et al. 2002), and pays special attention to the benefits people obtain from biodiversity, ecosystems and their functions. Following the MA, there are four types of services: 1) provisioning (products obtained from ecosystems), 2) regulating (e.g. flood control; species balance), 3) cultural (non-material benefits obtained from ecosystems), and 4) supporting, which maintain all other services by for example primary production and nutrient cycling (MA 2003). These services provide the basis for human well-being (e.g. MA 2003; Carpenter et al. 2009). Preserving the ability of ecosystems to produce services can be understood to form the basis for moral, intellectual and economic imperatives for environmental governance, which should aim to build mechanisms to conserve, keep up and improve ecosystems' services. This is an anthropomorphic perspective for understanding and relating to the environment (it is beyond the scope of this article for us to engage in a critique of this, however), but nevertheless, it offers a simplistic way for policy makers to conceptualise complex interrelationships between nature and society (Figure 1).

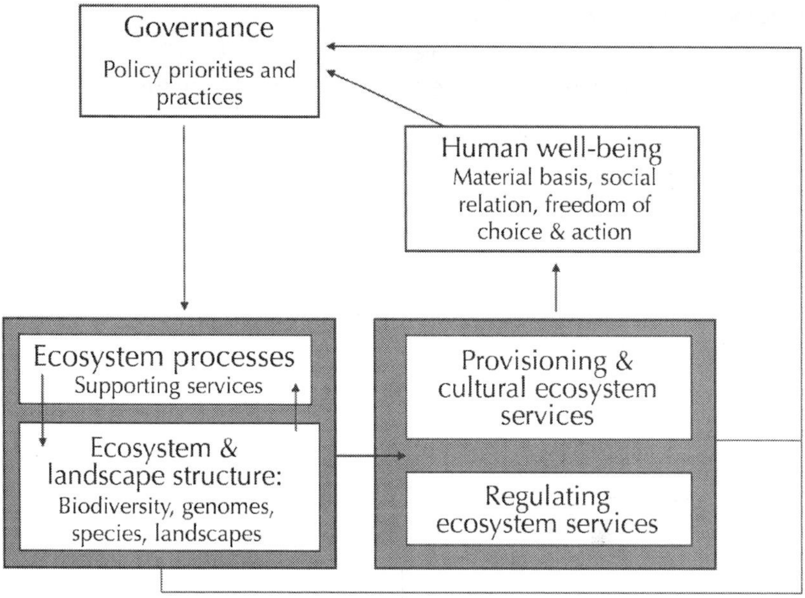

Figure 1: Ecosystem services, well-being, and governance (De Groot et al. 2002; MA 2005; Carpenter et al. 2009).

In order to make the concept of ecosystem services more concrete, some examples of ecosystem services are provided in Table 1; they relate to forest ecosystems and were chosen because of their links to our reindeer herding case. We do not provide indicators for supporting services, but just to the other three types of services, which are kinds of 'end services' and are sustained by supporting services. In addition, Table 2 outlines indicators used for assessing human well-being.

Table 1: Examples of ecosystem services (see Constanza et al. 1997; De Groot et al. 2002; MA 2005; Stenger et al. 2009) and indicators used in our case study

Ecosystem services	Indicators used in the case study
Supporting:	
Soil formation, primary production, nutrient cycling	

Regulating	
Population and keystone species control through trophic-dynamic relations, stabilising species balance.	-Balance between wolves, reindeer, wild forest reindeer, and moose.
Enhancing the reproduction of important species by suitable living and reproduction habitats.	-Estimation of the predator / wolf densities in the reindeer pastures and habitats.
	The availability of prey animals for predators.
Improving species resistance to disturbances e.g. by maintaining gene pools and viability of species populations.	-Ability of wolves to move across borders and exchange genes with different sub-populations.
	-Estimation whether the number of wolves will increase or decrease.
Provisioning	
Primary production of extractable, which are utilised in hunting, gathering, herding, berry picking etc.	-Estimation of number of reindeer that ecosystem sustains.
	-Estimation about the amount of lichen reindeer can utilise.
Sociocultural	-The availability of assumingly attractive landscapes and activities within those.
Environments and landscapes that support ecotourism, recreation and outdoor activities.	
Existence of utilisable or otherwise valued species.	-The availability of assumingly attractive species (e.g. predators, but also reindeer).

Heikkinen *et al.*

Heikkinen *et al. Pastoralism: Research, Policy and Practice* 2012 2:11, doi:10.1186/2041-7136-2-11

Table 2: Components of human well-being and their indicators (based on MA 2005)

Components of well-being	Indicators used in the case study
Freedom of choice	Participation vs. determining the option(s) (e.g. can herders choose to continue their livelihood).
Material basis for life	Possibilities for income for reindeer herders.
Social relations	Ability to continue traditional herding which is embedded with social practises; possible new social relations.

Heikkinen *et al.*

Heikkinen *et al. Pastoralism: Research, Policy and Practice* 2012 2:11, doi:10.1186/2041-7136-2-11

STUDY AREA: REINDEER HERDING AND NATURE CONSERVATION IN FINLAND

Finnish Reindeer Herding

Our case study of Finnish reindeer herding comes from the municipality of Kuusamo in the northeastern part of the Oulu province of Northern Ostrobothnia. It is a sparsely populated and largely forested area bordering Russia. In Kuusamo and neighbouring areas, reindeer herding has been a traditional livelihood since the seventeenth century when Finnish settlers adopted herding practises from indigenous Forest Sámi. Reindeer herding then spread deep into Russian Karelia until Finnish independence in 1917 and finally, the Second World War and its aftermath temporarily depopulated large parts of the Russian side of the border area. The loss of old pasture areas and the rapidly increasing industrial forestry activities forced reindeer herders to replace tree lichen pastures (which were particularly important in winter and spring) with hay feeding and reindeer farming (Kortesalmi

Figure 2: Koillismaa region, Kuusamo, Ruka ski resort and Oulanka National Park.

ONP is managed by Metsähallitus whose public administration duties include the management of protected areas (Heinonen 2007). The Natural Heritage Services of Metsähallitus, funded mostly by the state, manages Finnish national parks within the Parliamentary legislation, the Ministry of the Environment's guidelines (e.g. 2002) and Metsähallitus (e.g. 2000) own principles and management and land use plans for each park. Following the International Union for Conservation of Nature's (IUCN) protected area management categories, ONP is a category II (national park). Gilligan et al. 2005) stated that ONP had the highest management effectiveness score of any area assessed in a survey of 206 forest protected areas worldwide (Dudley et al. 2004).

ONP has become one of the most popular and well-known parks in Finland. In 2008, there were 163,000 visits to ONP, almost three times as many as in 1992. The joint goal of ONP and the adjacent Paanajärvi national park in Russia is to become the most attractive wilderness location in Europe by 2015 (Metsähallitus 2010). The aim is to achieve a competitive advantage over other national parks in Europe by combining sustainable tourism and nature conservation. According to a study conducted with use of the money generation model, park visitors boosted the local economy by 14.2 million Euros in 2008 (Metsähallitus and Metla [Finnish Forest Research Institute; Helsinki, Finland] 2009). Thirty partnership companies organise recreation services in ONP, around twenty accommodation companies are located close by, and thirteen businesses have acquired the PAN Parks partner certification in the Oulanka region.

ONP became certified as PAN Parks in 2002 (Zinke et al. 2002) and the adjacent Russian Paanajärvi National Park (104,371 ha) in 2005. Together, parks form a potential cross-border ecological corridor and hence, can be considered as a key area for advancing the idea of Fennosscandia Green Belt and, for example, viable and cross-border migratory great beast populations have a central role in this initiative (cf. Paanajärvi 2010; Metsähallitus 2010).

The PAN Parks Initiative

The PAN Parks initiative provides a third-party certification system under the World Commission on Protected Areas Framework for Management Effectiveness. PAN Parks was established in 1997 by the WWF and the Dutch leisure company Molecaten (Font and Clark 2007). A park becomes certified following verification carried out by a team of independent experts in accordance with the PAN Parks (2008) five principles, each with specific criteria: natural values, habitat management, visitor management, sustainable tourism development, and tourism business partners. The minimum size of a PAN Park is 20,000 ha with a wilderness/core zone of at least 10,000 ha in its natural state, which represents the most undisturbed expanses of Europe's remaining natural landscapes. The PAN Parks criteria do not permit extractive uses in the core zone (e.g. hunting/culling, fishing or motorised transportation) even if based on traditional use. PAN Parks exerts pressure towards certain management options through the verification process, but national legislation is complied with. The PAN Parks claims to combine wilderness protection and sustainable tourism, which 'provides real benefits for the rural communities in and around the protected areas' (PAN Parks 2010). However, as PAN Parks favours a non-intervention conservation approach, it also prefers exclusion of semi-domesticated reindeer; it has claimed that any impacts of semi-domesticated reindeer on ecosystems should resemble those of wild forest reindeer, which are defined as a 'natural species'. The main concern is the number of grazing reindeer within the park — a hypothetical number of wild forest reindeer is claimed to be lower than the current number of semi-domesticated reindeer (Jartti et al. 2010).

The possible benefits of PAN Parks certification for local social development via increasing tourism is still questionable, since only few tourists have used PAN Parks certification as a key motive to choose Oulanka as their travel destination (Ylimaunu 2009). However, PAN Parks may still provide better marketing networks for certified local tourism entrepreneurs. Yet, it seems that despite benefits that PAN Parks promises for local development, it benefits different local groups differently, and cannot be viewed as all-encompassing solutions to contradictions between conservation and development, not to mention dilemmas between conservation of wildlife and pastoral development (Sarkki et al. In press).

'The PAN Parks wilderness area in Oulanka is a unique example of meandering river ecosystem surrounded by untouched boreal forest near the Arctic Circle. It is the first certified piece of PAN Parks Wilderness in Finland with the ambition to protect fragments of unlogged boreal forest from intensive reindeer herding.' (PAN Parks 2010).

The Green Belt Initiative

The European Green Belt initiative is coordinated by the IUCN and aims to connect national parks, reserves and transboundary protected areas along and across national borders. The aim is to harmonise management methods on both sides of country boundaries. The Green Belt follows the border between eastern and western Europe as it was defined (and became commonly known as the Iron Curtain) at the end of the Second World War. The reason for following this line is that these areas have remained relatively undisturbed from human interventions and have potential to become natural corridor between national states as plenty of biodiversity values are found from these border areas, including predator populations (IUCN 2010a). According to IUCN:

'The Fennoscandian and Baltic States Green Belt includes some of the last remaining old growth boreal forests, – It is mainly due to the Cold War and the consequent low economic activities, that this area remains in a nearly pristine state. – The plant and animal world is unique due to the presence of endemic animal species such as taiga reindeer, grey wolf, brown bear, wolverine, ringed seal, land-locked salmon' (IUCN 2010b).

In a sense, the creation of the European Green Belt entails an erasure of the past, an obliteration of what was an often bitter, tense and violent European political and social history, in favour of the celebration of a natural history that, while appearing to be present in areas that are last vestiges of 'natural' landscapes, is nonetheless thriving in places that have been emptied of people, or where people are now considered to be intrusive.

Conservation Pressures on Reindeer Herding Wild Reindeer Effects

On several occasions, international and national conservation organisations have presented their arguments about what they see as the negative impacts of reindeer grazing on the 'natural state' of protected areas of northern Finland. For example, worries over the destruction of natural habitat by reindeer have been outlined in the IUCN evaluation report on Finnish protected areas (Gilligan et al. 2005), and IUCN's Finnish member also communicated this view in the annual reindeer herders' general meeting by stating that reindeer do not fit into national parks unless their effects on nature would resemble those of wild reindeer (Poropäivät 2005, cf. Nieminen 2008). Similar kinds of discussion about acceptable wild reindeer effects in conservation areas have focused for example, in the Malla strict nature reserve, classified as category IUCN I a in IUCN's protected area classification system, Pallas-Ylläs National Park (Jokinen 2005; Heikkinen et al. 2010), and were referred to during the PAN Parks certification process for Oulanka National Park (Sarkki et al.In press).

The wild reindeer preference is mentioned several times in ONP's verification renewal report (Stritih et al. 2007). Recently, a more modest plan was signed, according to which park management commits to reduce reindeer grazing within the park over the next few years (PAN Parks 2010). Something particularly problematic from a reindeer herding perspective, as we have found from interviewing herders, is the very low estimations of the natural wild forest reindeer densities which would not support any sort of livelihood without a considerable reformulation of the herding profession. Furthermore, motorised cross country transportation, particularly during the winter time, is an essential part of reindeer herding livelihoods today, but it is also obviously contradictory to the wild reindeer preferences. Similarly, different kinds of fence systems, also proposed by PAN Parks 2002 (Zinke et al. 2002), are problematic except those that are needed to control reindeer movements for culling or earmarking purposes. The leader of the Ala-Kitka reindeer herding cooperative stated in one interview: 'We have the right to free grazing and it is not reasonable for some areas to be fenced'. Furthermore, he pointed out to us that ONP has been verified as PAN Parks with the decisions of only a few people, and this new status turned reindeer, 'which have been here for ages', into a threat overnight.

Viable Predator Populations

The populations of what in Finland are called 'great beasts' are a key issue in many major nature conservation initiatives, for example as in the IUCN's (2010b) European Green Belt plan, and were recognised as such by PAN Parks during the ONP certification process, even though PAN Parks finally concluded that wolves in particular are problematic in ONP in accordance with current reindeer traditions. The recommendations later changed to allow for consideration of possible culling arrangements for special problem animals (e.g., Miller 2005). However, the pressures for further predator hunting limitations are on the agenda and are focused specifically on prohibiting the hunting of bears in the core zone of ONP (Miller 2005, Stritih et al. 2007; cf. Poropäivät 2005; Nieminen 2008).

One reindeer herder described an incident involving a bear inside his colleague's calving fences located near the ONP:

"'He saw a bear chasing a calf and tried to disturb the bear with his all-terrain vehicle, but the bear just looked at him, took the calf and went away. The bear had killed 12 calves during the night'. Secondly, reindeer must be taken inside winter fences early because the predation risk and feeding costs have increased. A herder living next to the park said: 'There are good pastures in the park. ... I would take none of my reindeer home, but predators force me to do it, and they are increasing'."

According to our previous study, it is possible to reconcile the presence of other great beasts, such as lynx, bears and wolverines, with traditional reindeer herding — with the one exception of the wolf, which 'stirs the whole system' (Heikkinen et al. 2011). In 2008 and 2009, during our interviews with local reindeer herders and agricultural officers, who are responsible for animal-based compensations, all emphasised that reindeer kills by wolves challenges the viability of local livelihoods based on herding. For example, in the municipality of Pudasjärvi in 2007, a wolf pair killed or ripped some hundred reindeer in a period of six months and the work time of local herders changed into frustrating conservation hunting efforts and into the search for dead reindeer, as well as those reindeer that had escaped from their own pastures to those of neighbouring reindeer cooperatives (Heikkinen et al. 2011). An important suggestion made by the herders on that occasion was that the conservation hunting and predator population management costs should be fully compensated, or at least paid and arranged by those organisations that propose predator conservation.

NATURE CONSERVATION AND REINDEER HERDING: FOUR SCENARIOS FOR THE FUTURE

We now move on to the construction of four scenarios that are based on different policy priorities related to nature conservation (Figure 3). Before each scenario, some necessary background is presented that examines existing Finnish and international developments that support the given scenario. The four scenarios on reindeer herding and nature conservation are called: A) adapting traditional herding, B) reindeer farming, C) compensated ecosystem service producer, and D) transformed professions. We structure these scenarios according to the following dimensions: 1) policy priorities, and practises 2) regulating services, 3) provisioning and cultural services, and 4) reindeer herders' well-being.

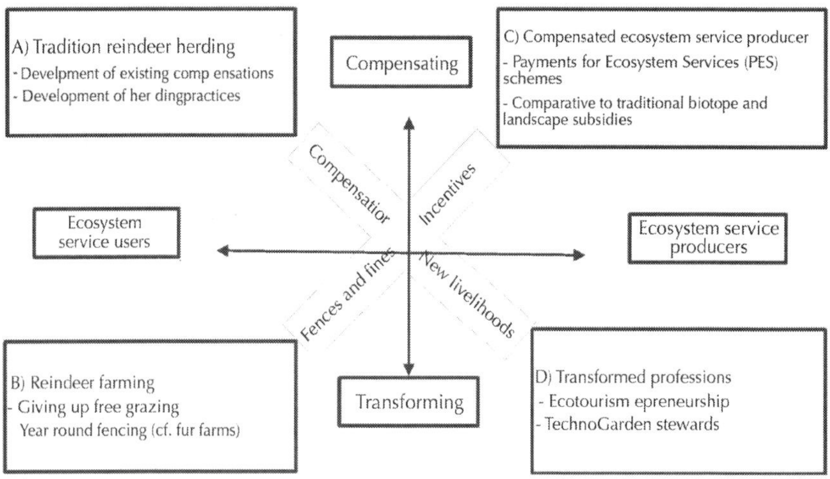

Figure 3: Four development paths for reindeer herding.

For reindeer herding, the most important policy choices will be made between enhancing compensation systems or transforming livelihood and defining herding either as a user or as a producer of ecosystem services. If defined as a user, we assume that development might follow either A) adapting current herding and compensation methods or B) increasing regulating policies to give up free grazing and transforming free grazing reindeer herding into a direction of year-round farming (cf. Heikkinen et al. 2007). In Scenario A, compensations are paid if certain management decisions, like predator conservation, have negative impacts on reindeer herding. In Scenario B, herding is merely seen as a livelihood using and deteriorating ecosystem services and hence efforts are made to halt negative effects of herding on biodiversity and ecosystem services.

If reindeer herding will be defined as an ecosystem service producer, we assume that some innovative solutions might unfold. For example, reindeer herders could be incentivised for producing ecosystem services that would have marketing value in ecotourism and simultaneous value for ecological functions. If, following the trend on developing incentive schemes, we assume that Scenario C) certain systems of payments for ecosystem services might develop. Instead, if nature conservation policies will develop towards creating strictly interpreted 'wildernesses' and excluding traditional engagements, we assume that some more elementary transformations in reindeer herding will follow. Here we assume that transformations would particularly serve the needs of both ecotourism and ecosystem management purposes, and for herders, these could materialise in Scenario D) as some sort of tourism entrepreneurship or park stewardship professions.

Scenario A) Adapting Traditional Reindeer Herding

Compensation programmes are already an approach widely used for integrating local human interests and environmental protection measures. For example, compensation has traditionally been paid to cover land owners' economic losses for certain developments, such as expropriating lands for conservation areas (Ruuttula-Vasari and Juvonen 2006). In reference to our case, the compensation system for predator damages in relation to Finnish reindeer herding is twofold. Firstly, the land-beast compensation system is based on found and proven damages to the individual reindeer herders (Sippola et al. 2005; O Moilanen, unpublished thesis; Ollila 2008). Secondly, the change in the compensation system regarding the golden eagle has been quite successful. Different Reindeer Herding Cooperatives are currently being compensated, not based on found and proven damages, but in relation to the number of nesting eagles within the cooperative as well as on the basis of the assumptions about how large a portion of the eagles' diet is based on reindeer in a given area. Currently, herders themselves are informing the Ministry about eagle nests they have located, in order to receive compensation. With terrestrial predators, there are constant accusations these are being poached by herders without licences. In Sweden, an 'area-based' compensation scheme is also applied to terrestrial predators, but it has received some criticism because the compensation is not paid directly to the herder who suffered the costs of conservation, but to the Sámi Parliament (as only Sámi are allowed to herd reindeer in Sweden, unlike in Finland), which then chooses how the money is distributed between different localities that have been affected by predators (Naskali et al. 2006; Heikkinen et al. 2011).

Policy Goal and Practices

The policy goal is to enable the coexistence of reindeer herding and conservation priorities. Application of territory-based land-beast compensation system for reindeer herders works better than the current system based on found carcasses. In addition, electronic devices for finding carcasses and developing predator repellents are coming into use. The lost income of herders is compensated due to predation and diminished reindeer densities.

Regulating Services

The number of wolves and other predators has increased slightly in reindeer pastures and habitats. This has resulted in a decrease of reindeer numbers closer to the hypothetical number of wild forest reindeer, and in a slight increase of lichens. Wolves and other predators have increased because of hunting bans. Wandering great beasts strengthen the gene pool of a wider area but this serves the general targets of Green Belt and ecological corridor only partially, and the exchange of genes between Finnish and Russian wolf populations has not increased.

Provisioning and Cultural Services

The ecosystems can sustain fewer reindeer than before because of increasing numbers of predators. However, reindeer herding communities and traditional livelihood practises still exist, even in an adapted manner. Free grazing reindeer herding is accepted and used in tourism marketing. Moreover, predators are utilised in marketing the tourist destination, and the predator watching industry has slightly increased.

Human Well-being

Reindeer herders do not have any impact on the increasing number of predators. However, improved compensation systems ensure herders' income. Tourism serves only some extra markets for expensive reindeer products. Improved compensations also improve the herders' acceptance towards predators in the area. As free grazing herding still continues in an adapted manner, the community practises and associated social relations remain. Reindeer herding traditions and vocabulary will survive and adapt, and local interaction with nature will be intense. ONP serves a rather wide variety of societal targets, but spreading predator populations create anxiety in neighbouring communities.

Scenario B) Reindeer Farming

Perhaps the oldest way to resolve the tension between conservation targets and human interests is the so called 'Yellowstone model' or 'fences and fines' approach. It encompasses conservation measures by prohibitions, top-down policy control and fines, and even the very separation of human culture and society from conservation areas by fencing (Spence 1999; cf. Rytteri and Puhakka 2009). In Finnish reindeer discourses, not only the enclosure of reindeer outside certain areas, such as private forests, fields and conservation areas but also the ending of the free grazing right and the transformation of reindeer herding to farming have a long history of being discussed. Recently, the cessation of free grazing rights has been justified by references to, for example, the avoidance of traffic accidents, overgrazing problems and for nurturing larger, great beast populations (cf. Heikkinen 2002; Jokinen 2005; Rantamäki-Lahtinen 2008). Here reindeer herders are seen more strongly as users of ecosystem services, and an idea in conservation is that reindeer herders should not be allowed to deteriorate the forest floor and regional biodiversity. In addition, suggestions have been made for reintroducing wild forest reindeer to the area. There are reflections of this kind of trend in PAN Parks' recommendations to decrease the density of reindeer close to the hypothetical density of wild forest reindeer (Zinke et al. 2002, Stritih et al. 2007, PAN Parks 2010; Heikkinen et al. 2010; Sarkki et al. In press). An increase in predator numbers also results in difficulties for continuing free grazing reindeer herding. The number of wolves in Finland has also been debated between EU and Finland, as EU claimed in 2005 that Finland was violating the conservation targets of EU's habitat directive. After intense debate, the appeal was closed in 2008 when the European Commission finally stated that Finland has not threatened the sustainable level of its wolf population (Heikkinen et al. 2011; Hiedanpää and Bromley 2011). However, this debate stresses that international conservation pressures exist for increased predator conservation.

Policy Goal and Practices

The policy goal of this scenario is that natural biodiversity is a priority. Free grazing right is ended in national parks and the remaining herders feed and protect their animals inside extensive fence systems in their private lands, and their livelihood is part of normal agricultural policies and subsidies. A wild forest reindeer population is reintroduced and kept up in ONP, to resemble the pristine boreal forest ecosystem.

Regulating Services

Predator numbers have increased but are limited because the availability of prey, the wild forest reindeer. Great beasts are intermittently present and follow the free movements of wild forests reindeer. Wandering great beasts strengthens the gene pool of the wider area and this serves the general targets of the Green Belt and ecological corridor targets. Lichens and perennial plants are abundant and sustain the hypothetical natural number of wild forest reindeer. Sparse wild forest reindeer populations have only minor effects on plant cover. All movement tracks are light, and also tourists' presence in ONP is limited.

Provisioning and Cultural Services

As free grazing reindeer herding has ended, the ecosystems are not anymore serving to sustain semi-domesticated reindeer populations. Wild forest reindeer are not hunted. However, hay farming has increased among reindeer farmers, in order to decrease their costs of fodder. Recreational and tourism services of ONP have decreased, but the value for science and educational purposes has increased because of the presence of natural animal populations. Some of the tourists favour 'natural dynamics', but others fear unmanaged and non-hunted great beast populations.

Human Well-being

Traditional herding practises and communities are lost, and herders do not have the freedom to choose whether they will continue their traditional livelihood. Reindeer-related traditions, culture and language are conserved in archives and museums. Local interaction with nature is sparse, and social relations embedded in free grazing reindeer herding practises are changed. The herders' income comes from reindeer farming, which is subsidised like sheep farming, for example. Tourism, park management and remaining reindeer farming are separate engagements. Particularly problematic for neighbouring areas are the effects on wild animals followed by great beasts and the fluctuating wild animal populations. Also neighbouring Reindeer Herding Cooperatives increasingly give up free grazing.

Scenario C) Compensated Ecosystem Service Producer with Pes Schemes

The third conservation trend develops in direct connection to the ecosystem services approach: different PES have been developed for enhancing ecosystem service production (Jack et al. 2008; Sommerville et al. 2009), and PES can be used to balance the inequalities in distributing costs and benefits of proposed conservation measures (Bruner et al. 2008). In short, PES systems are built to make conservation economically and locally attractive (cf. Daily and Matson 2008; Pejchar et al.2007). A combined approach of conservation and development may offer simultaneous environmental and economic advantages locally by incentivising local engagements towards park keeping, conservation hunting, meadow mowing or sheep grazing services. In fact, the EU CAP subsidies and other agro-environmental schemes of the EU for keeping up traditional landscapes and biotopes aleady follow this logic (CAP 20102011; Birge and Fred 2011 unpublished work). Regarding other sectors in Finland, Finnish forestry in northern state lands is increasingly including 'neoliberal' elements, and market-based governance seems often to over-run previously state-based decision making (Sarkki and RÃ¶nkÃ¤ 2012). In addition, a PES scheme called 'nature values trade' regarding private forests has been a rather successful experiment in efforts to combine forest ownership and nature conservation (Paloniemi and Vilja 2009). A key difference with the two previous scenarios (A and B) is that here reindeer herders are viewed as producers of ecosystem services and are incentivised for doing so.

Policy Goal and Practices

The policy goal of this scenario is to promote synergies between business and conservation. Biodiversity and local socio-cultural priorities are guaranteed with a wide variety of ecolabels, which are utilised in marketing 'sustainable nature tourism'. Reindeer herders are supported for producing certain ecosystem services. Particularly, PES are paid for creating wild reindeer simulating grazing effects and for allowing predation on reindeer. Herders are incentivised for keeping up some reindeer population year round inside ONP for predators and tourists. PES are paid also for conservation hunting and maintaining stable great beast populations and preventing their spread to the neighbouring areas.

Regulating Services

Carnivore-herbivore natural dynamics are working partially, but population fluctuations and movements are stabilised by human interventions (e.g. by conservation hunting and controlled number and movements of reindeer). Large predators have only a slight effect on surrounding areas and hence only partially support the general targets of Green Belt and ecological corridor ideologies. There is no exchange of genes between predator populations, because predator movements are highly managed. However, some new genes may be imported from other populations. Semi-domesticated reindeer simulate wild reindeer in the ONP ecosystem, and thus the number of reindeer is kept quite small within the park.

Provisioning and Cultural Services

As reindeer herding is centred on substituting wild forest reindeer, especially within the park, the number of animals decreases. The production of meat and compensation form the basis of reindeer herder incomes. Predators are present and utilised in the tourism industry (e.g. such as bear watching). Predator-watching enterprises pay compensation to herders. Ecolabel boosted tourism creates local income. The value of reindeer products is very low and serves tourism directly, but mostly in marketing. Tourism routes are clear and build to sustain rather a great number of visitors.

Human Well-being

Nature-based tourism combined with compensated ecosystem service production guarantees higher income for local herding communities, but parts of the traditional herding-related practices have disappeared. Herders can choose if they take part in incentivised practises, but those reindeer herders who take part in incentivised ecosystem service production have better income possibilities than those who continue with more traditional herding. Part of the reindeer-related cultural and linguistic traits will disappear, but local interaction with nature is widespread and regular. Herding is still a separate profession, but amalgamated partially to nature-based tourism and park management activities.

Scenario D) Transformed Professions

Here reindeer herders have two opportunities for transforming their profession: herders could become so called 'ecosystem stewards' or transform their livelihoods into ecotourism entrepreneurship. Ecotourism has been recognised as one of the most important ecosystem services provided by protected areas (Kettunen et al. 2009). Reindeer herders are already assumed and encouraged to benefit from increasing ecotourism through transformations from 'extractive' livelihoods to ecotourism-oriented businesses (Stritih et al. 2007). Local park stewardship is another potential for the future, also in line with the idea of an environmental engineering approach put forward by MAs (2005) TechnoGarden scenario. Ecosystem stewards are those 'whose actions modify the quantity or quality of ecosystem services available to environmental service beneficiaries. Ecosystem stewards are recognised by society as having the right to interact with an ecosystem, provided they accept limitations on those rights and obligations to maintain the ecosystem' (Swallow et al. 2009). In short, the local educated stewards are paid for producing the services, and their salaries will compensate the lost income possibilities from traditional livelihoods and changes in socio-cultural practises. The role of locals as ecosystem stewards has been proposed for example in relation to hunting of predators (Holsman 2000). It has been also suggested that wolves and other predators could keep the herbivore dynamics stable within national parks (Brown 2010). It would be then the task of human stewards to regulate the predator populations.

Policy Goal and Practices

The policy goal includes production of ecosystem services by widespread environmental engineering. Semi-domesticated reindeer are used as substitutes for wild forest reindeer, and managed as prey for predators. Biodiversity-concentrated ecolabels cease free grazing rights inside the ONP, and former herders are encouraged and subsidised for ecotourism-related businesses or for taking jobs as Park stewards, and become normal paid employees of Park management. Adoption of 'zoo-like' management practises and widespread marketing of 'the Wilderness Experience Park' are taking place.

Regulating Services

ONP is managed to support and conserve a high variety of endangered species and habitats inside the park. It serves other ecosystems mostly as a gene bank, and genetic exchange between Finnish and Russian wolf populations is ensured by increasing collaboration in transboundary Oulanka-Paanajärvi Park. For supporting the highest possible variety of ecosystem functions, ONP's nature is managed in a zoo-like manner. For example, animal populations are highly managed, their movements are controlled by fencing, and populations are kept stable by conservation hunting and regular reintroductions.

Provisioning and Cultural Services

There would not be any reindeer meat production for human consumption, as in the previous scenarios, and economic values are created only through ecotourism. High management level makes high tourism numbers and new programmes possible. For example, ecotourism entrepreneurs sell predator-watching programmes. Park stewards control, manage and create habitats inside ONP and keep populations stable and control animal movements outside of the park. Park stewards can also make conservation hunting part of tourism business, by involving customers in arranged activities. Tourism infrastructures are hidden and routes lead to various wilderness spotting places.

Human Well-being

Local traditional livelihoods are lost and newly transformed professional roles guarantee high local income only for those who are able to adjust to them. Those who are engaged in these transformed professional roles interact with nature intensively and are part of the process of creating a 'Wilderness Experience Park' either as ecosystem stewards or ecotourism entrepreneurs.

CONCLUSIONS: UNCERTAINTIES IN THE ECOSYSTEM SERVICE APPROACH AND SCENARIOS

Our scenarios show that there are many compatibilities as well as incompatibilities between pastoralism and nature conservation, and there are multiple ways by which the ecosystem service framework could be applied (Table 3). We conclude that even though seeing reindeer herders as producers of ecosystem services may offer some innovative solutions to dilemmas between pastoralism and conservation, there are also various uncertainties related to the ecosystem service approach. The uncertainties involved are firstly related to a number of questions, such as: in what ways can human-created services substitute 'natural' ecosystem functions? What degree of ecological complexity is needed for service production? What are the limits and indicators of successful environmental engineering in large areas (see MA 2005)? It seems evident that even rather light environmental engineering does change the behaviour of wildlife, but how and to what extent does this lead to genetic variation and change?

Table 3: Summary of tentative trade-offs related to scenarios

Scenario A) Traditional reindeer herding	Scenario C) Compensated ecosystem service producer
Gainers:	Gainers:
-Pastoralist's home economics	-Manageability of predator and semi-domesticated reindeer
-Predator populations (easy prey)	-Continuing local herding communities
-Semi-domesticated reindeer's instincts and survivability (continuous predation pressure)	-Incentivized motive for reindeer management
	-Natural grazing pressure simulations
Losers:	Losers:
-State (high compensations costs)	-State (high costs of incentives)
-Traditional food for humans production motive	-Predator watching enterprises (payments for herders for increasing predation)
-Motive for reindeer management (changes to searching of killed reindeer)	-Traditional food for humans production motive
-Predators hunting instinct and skills	-Natural dynamics of hunting pressure

Scenario B) Reindeer farming	Scenario D) Transformed professions
Gainers:	Gainers:
-Traditional food for humans production motive	-Manageability, stability and commodification of local environment
-Agricultural investment economy (rising costs of production, but steady harvest)	-Wild forest reindeer population
	-Local and semi-local conformists (educated, probably young opportunists)
	-Tourism
	-Municipal economies
Losers:	Losers:
-Domesticated reindeer meat quality	-Traditionalists (local conservatives not willing for modern occupations)
-Domesticated reindeer's instincts and survivability	-Nature's uncontrolled dynamics
-Predator populations (loss of easy prey)	-Traditional local culture and communities
-Reindeer herders' socio-cultural traditions and freedom of choice regarding their livelihood.	

Heikkinen *et al.*

Heikkinen *et al. Pastoralism: Research, Policy and Practice* 2012 2:11, doi:10.1186/2041-7136-2-11

The second set of uncertainties is related to more general assumptions about ecotourism and cultural services. Tourism could benefit from predators, since predator-watching is an increasingly significant business in Finland, but would the engineered 'zoo-like' parks satisfy tourists' quest for a wilderness experience? Or would the increased number of predators create fear in tourists?

Thirdly, despite the anthropocentric dimension of the ecosystem service concept, the question remains of who actually benefits from the ecosystem service production approach in nature conservation. There is a relative lack of research on the interrelations between ecosystem services production and human well-being (Carpenter et al. 2009), and the link between local community well-being and ecosystem service production can be, and often is, negative. In addition, it has been argued that the concept of ecosystem services in fact blinds us from understanding ecological, economic and social complexities and interrelationships (Norgaard 2010). This blinding might take place, for example, when applying ecosystem service-based incentives to local resource governance. Can these monetary measures also compensate for the social and cultural changes taking place with the new governance tools? Will pastoralists accept the production of livestock mainly for predators? It seems that many conservation organisations have adopted popular scientific concepts and use them as justification for conservation targets. The danger is that the ecosystem service approach remains as distant to local well-being and experience as have the earlier 'fences and fines' conservation approaches or species conservation agendas. Hence, pastoral communities should be engaged in the discussions over nature conservation, so as to establish proper links between ecosystem services and community well-being.

The third uncertainty is related to PES schemes. It has been documented that PES schemes may produce negative impacts on local livelihoods and that their effects are more complex than simple assumptions of local benefits (e.g. Grieg-Gran et al. 2005). In the light of our scenario exercise, PES schemes may in fact produce new kinds of inequalities at the local level between those who are able and willing to transform their practises guided by such schemes, and those who will not or cannot transform their practises. Thus, even though PES schemes seem to be promising for herders, in comparison to other scenarios, PES schemes may generate new kind of inequalities and contribute to social tensions at the local level.

The discussions between pastoralists and conservation agencies could be aided by scenario exercises, such as the one presented in this article. Participatory workshops could be arranged by researchers using scenarios on the relationships between pastoralists and nature conservation agencies, as a starting point for discussions. In this way, local values, understandings and every day practises could be evaluated in relation to conservation targets. Whatever the final policy solutions will encompass in future, their success will be bound to and legitimated by local commitment and acceptance. Without this, what are intended as policy practises might end up as mere policy exercises.

AUTHORS' CONTRIBUTIONS

Empirical studies for this article were conducted primarily by HH and SS. HH focused more on predator and reindeer problematics while SS studied the relationships between reindeer herding and Oulanka National Park. Analysis of the research material and developing the scenario exercise was a joint intellectual effort of HH, SS and MN. Actual article writing was equally performed by HH and SS, and MN provided international expertise and points of comparison for the Finnish case as well as editorial help. All authors read and approved the final manuscript.

ACKNOWLEDGMENTS

We like to thank reviewers and editors for constructive comments. This study was conducted as part of the Finland Distinguished Professor (FiDiPro) programme 'Human–environment relations in the north: resource development, climate change and resilience' funded by the Academy of Finland and based at the Thule Institute, University of Oulu, Finland and the Department of Anthropology, University of Alberta, Canada. Some funding was also received from the Emil Aaltonen Foundation.

REFERENCES

1. Birge, T, and M Fred. 2011. (Unpublished). Using Social-Ecological Inventory to Identify Ecosystem Services and Stewards of Traditional Rural Biotopes in SW Finland. http://www.bioecon.ucl.ac.uk/12th_2010/Birge.pdf . Accessed 29 September 2011
2. Bisi, J, S Kurki, M Svensberg, and T Liukkonen. 2007. Human dimensions of wolf (Canis lupus) conflicts in Finland. European Journal of Wildlife Research 53 (4): 304–314.
3. Brown, M. 2010. Ranger Wolf: predators pushed as park stewards. The Seattle Times. 7 February 2010. http://seattletimes.nwsource.com/html/localnews/2011008041_apmtwolfnation1stldwritethru.html. Accessed 29 September 2011

4. Bruner, A, R Naidoo, and A Balmford. 2008. Review of the costs of conservation and priorities for action. http://ec.europa.eu/environment/nature/biodiversity/economics/teeb_en.htm . Accessed 15 February 2010
5. Buller, H. 2008. Safe from the wolf: biosecurity, biodiversity, and competing philosophies of nature. Environmental Planning A 40: 1583–1597.
6. CAP. 2010. Agriculture and rural development. http://ec.europa.eu/agriculture/envir/ . Accessed 2 March 2010
7. CAP. 2011. Agriculture and rural development: taking care of countryside. http://ec.europa.eu/agriculture/capexplained/countryside/index_en.htm . Accessed 29 September 2011
8. Carpenter, S, H Mooney, J Agard, D Capistrano, R DeFries, S Díaz, T Dietz, A Duraiappah, A Oteng-Yeboah, H Pereira, C Perrings, W Reid, J Sarukhan, R Scholes, and A Whyte. 2009. Science for managing ecosystem services: beyond the Millennium Ecosystem Assessment. Proceedings of the National Academy of Sciences of the United States of America 106 (5): 1305–1312.
9. Chan, K, R Pringle, J Ranganathan, C Boggs, Y Chang, P Ehrlich, P Haff, N Heller, K Al-Khafaji, and D Macmynowski. 2007. When agendas collide: human welfare and biological conservation. Conservation Biology 21 (1): 590–68.
10. Convention on Biological Diversity (CBD). 2010. Convention on biological diversity: revising and updating the CBD strategic plan, post-2010. http://www.cbd.int/sp/sp2010+/ . Accessed 15 February 2010
11. Costanza, R, R d'Arge, R de Groot, S Farber, M Grasso, B Hannon, S Naeem, K Limburg, J Paruelo, RV O'Neill, RG Raskin, P Sutton, and M van den Belt. 1997. The value of the world's ecosystem services and natural capital. Nature 387: 253–260.
12. Countdown. 2010. Save biodiversity, move from words to action. IUCN. http://www.countdown2010.net/biodiversity Accessed 28 September 2011
13. Daily, G. 1997. Nature's services: societal dependence on natural ecosystems. Island Press, Washington.
14. Daily, G, and P Matson. 2008. Ecosystem services: from theory to implementation. Proceedings of the National Academy of Sciences of the United States of America 105 (28): 9455–9456.

15. De Groot, S, M Wilson, and R Boumans. 2002. A typology for the classification, description and valuation of ecosystem functions, goods and services. Ecological Economics 41 (3): 393–408.
16. Dudley, N, A Belokurov, O Borodin, L Higgins-Zogib, M Hockings, L Lacerda, and S Stulton. 2004. Are protected areas working? An analysis of forest protected areas by WWF. WWF International, Gland.
17. 2009. U Nature & Biodiversity Partnerships. http://ec.europa.eu/environment/nature/partnerships/ . Accessed 17 February 2010
18. Farber, S, R Costanza, and M Wilson. 2002. Economic and ecological concepts for valuing ecosystem services. Ecological Economics 41 (3): 375–392.
19. Finnish Nature Conservation Act. 1996. 1096.
20. Font, X, and S Clark. 2007. Certification of protected areas: the case of PAN Parks in Europe. In Quality control and certification in ecotourism, ed. Black R, Crabtree A, 299–315. CABI, Wallingford.
21. Gallopín, G, A Hammond, P Raskin, and R Swart. 1997. Branch points: global scenarios and human choice. Stockholm Environment Institute, Stockholm, Sweden.
22. GBO 2. 2006. Summary of the Second Global Biodiversity Outlook. Convention on biological diversity. http://www.cbd.int/gbo/ . Accessed 17 February 2010
23. Gilligan, B, N Dudley, A de Tejada, and H Toivonen. 2005. Management effectiveness evaluation of Finland's protected areas. Nature Protection Publications of Metsähallitus, Series A 147. http://www.metsa.fi/MEE . Accessed 11 January 2008
24. Grieg-Gran, M, I Porras, and S Wunder. 2005. How can market mechanisms for forest environmental services help the poor? Preliminary lessons from Latin America. World Development 33 (9): 1511–1527.
25. Heikkinen, H. 2002. Sopeutumisen mallit — Poronhoidon adaptaatio jälkiteolliseen toimintaympäristöön Suomen läntisellä poronhoitoalueella 1980–2000. [Integrating models adaptation of reindeer herding post-industrial environment in the western Finnish reindeer herding area 1980–2000] Hakapaino: PhD Dissertation University of Oulu. Finnish Literature Society.

26. Heikkinen, H, J Hukkinen, O Jääskö, A Laakso, L Müller-Wille, S Nevalainen, K Raitio, and N West. 2003. Poronhoidon tulevaisuus. Raportti EU: RENMAN-hank keen Kittling työpajasta 13–15 August 2003. [The future of reindeer management: report of the EU RENMAN Project Workshop in Kittilä 13–15 August 2003] Technology, Society, Environment July 2003. Helsinki University of Technology Laboratory of Environmental Protection, Espoo.
27. Heikkinen, H. 2006. Neo-entrepreneurship as an adaptation model of reindeer herding in Finland. Nomadic Peoples 10 (2): 187–208.
28. Heikkinen, H, S Lakomäki, and J Baldridge. 2007. The dimensions of sustainability and the neo-entrepreneurial adaptation strategies in reindeer herding in Finland. Journal of Ecological Anthropology 11: 25–42.
29. Heikkinen, H, O Moilanen, M Nuttall, and S Sarkki. 2011. Managing predators, managing reindeer: contested conceptions of predator policies in Finland's southeast reindeer herding area. Polar Record 47 (242): 218–230.
30. Heikkinen, H, S Sarkki, M Jokinen, and D Fornander. 2010. Global area conservation ideals versus the local realities of Reindeer Herding in Northernmost Finland. International Journal of Business and Globalisation 4 (2): 110–130.
31. Heinonen, M. 2007. State of the parks in Finland: Finnish protected areas and their management from 2000 to 2005. Nature Protection Publications of Metsähallitus A 170. Vantaa: Metsähallitus. Available at:http://www.metsa.fi/SIVUSTOT/METSA/EN/NATURALHERITAGE/PROTECTEDAREAS/STATEOFTHEPARKSREPORTING/Sivut/StateoftheParksReporting.aspx Accessed 17 August 2012
32. Hiedanpää, J, and D Bromley. 2011. The harmonization game: reasons and rules in European biodiversity policy. Environmental Policy and Governance 21 (2): 99–111.
33. Hirschnitz-Garbers, M, and S Stoll-Kleemann. 2011. Opportunities and barriers in the implementation of protected area management: a qualitative meta-analysis of case studies from European protected areas. Geographical Journal 177 (4): 321–334

34. Holsman, R. 2000. Goodwill hunting? Exploring the role of hunters as ecosystem stewards. Wildlife Society Bulletin 28 (4): 808–816.
35. Hukkinen, J. 2008. Sustainability networks: cognitive tools for expert collaboration in social-ecological systems. Routledge, London.
36. IUCN. 2010a. Broadening boundaries — naturally, the Green Belt of Fennoscandia. http://www.iucn.org/knowledge/news/?4898/ Broadening-boundaries — naturally. 11 March 2010. Accessed 18 August 2011
37. IUCN. 2010b. The European Green Belt Initiative. http:// europeangreenbelt.org/001.route_fs.html . Accessed 24 February 2010
38. Jack, B, C Kousky, and K Sims. 2008. Designing payments for ecosystem services: lessons from previous experience with incentive-based mechanisms. Proceedings of the National Academy of Sciences of the United States of America 105 (28): 9465–9470
39. Jartti, L, K Lahti, and M Hovi. 2010. Jartti et al. wilderness zone program for Oulanka National Park 2010. Jartti et al..
40. Jokinen, M. 2005. Poronhoidon ja suojelun vaikutukset Mallan luonnonpuistossa. METLA Kolari, 941. [Reindeer herding and protection effects Malla Strict Nature Reserve] Metsäntutkimuslaitoksen tiedonantoja. Metla Kolari.
41. Kettunen, M, S Bassi, S Gantioler, and P ten Brink. 2009. Assessing socio-economic benefits of Natura 2000 — a toolkit for practitioners. Institute for European Environmental Policy, Brussels, Belgium.
42. Kok, K, R Biggs, and M Zurek. 2007. Methods for developing multiscale participatory scenarios: insights from southern Africa and Europe. Ecology and Society 13 (1): 8.
43. Kortesalmi, J. 2008. Poronhoidon synty ja kehitys Suomessa. [Reindeer herding the birth and development in Finland] Helsinki: SKS.
44. Kortesalmi, J. 1996. Pohjois-Vienan Poronhoito. Talonpoikien poronhoidon alue, ominaislaatu, ikä, alkuperä ja kehityslinjat vuoteen 1922. [North of the white reindeer husbandry: the

peasants of reindeer herding area, character, age, origin and trends to 1922] Helsinki: Suomen muinaismuistoyhdistys [Doctoral Dissertation. The Finnish Antiquarian Society, University of Helsinki, Finland.

45. Mikkola, K, and E Mattila. 2009. Poronhoitoalueen etelä- ja keskiosien talvilaitumet. Tila paliskunnissa 2000-luvun alkuvuosina ja eräiden ravintokasvien esiintymisrunsauden muutokset merkkipiireissä 1970-luvulta lähtien. [The reindeer herding area in southern and central parts of the winter pastures. Status herding districts the 2000s and the early years of some food plants esiintymisrunsauden changes herald circles since the 1970s.]. 57. Metlan työraportteja [Finland: Metla Working Papers/Working Papers of the Finnish Forest Research Institute] 115, Finland.

46. Metsähallitus. 2000. The principles of protected area management in Finland. Guidelines on the aims, function and management of state-owned protected areas. Nature Protection Publications of the Finnish Forest and Park Service B 54. http://julkaisut.metsa.fi/julkaisut/pdf/luo/b054.pdf . Accessed 10 September 2008

47. Metsähallitus. 2010. Life Nature Projects and Structural Fund Projects. http://www.metsa.fi/sivustot/metsa/en/Projects/ . Accessed 2 March 2010

48. Metsähallitus and Metla. 2009. Kansallispuistojen ja retkeilyalueiden kävijöiden rahankäytön paikallistaloudelliset vaikutukset. [National parks and camping areas important to the local economy] Metsähallitus 2009. http://www.metsa.fi/sivustot/metsa/fi/Eraasiatjaretkeily/Virkistyskaytonsuunnittelu/suojelualueidenmerkityspaikallistaloudelle/Sivut/Kansallispuistoihinsijoitetutrahatpalautuvatmonikertaisina.aspx. Accessed 30 November 2009

49. Millennium Ecosystem Assessment. 2003. Ecosystems and human well-being: a framework for assessment. Island Press, Washington.

50. Millennium Ecosystem Assessment. 2005. Ecosystems and human well-being: synthesis. Island Press, Washington.

51. Miller, G. 2005. PAN Parks 2005: PAN Parks Verification. Oulanka National Park, Finland. Monitoring Report. PAN Parks.

52. Ministry of the Environment. 2002. Ohjelma luonnon virkistyskäytön ja luontomatkailun kehittämiseksi. The Finnish Environment 535. [SY535 Program of the recreational use and nature-based tourism development]. Finnish Environment Institute, Finland.

53. Naskali, A, H Juha, and S Leila. 2006. Biologinen monimuotoisuus talouskysymyksenä. Suomen ympäristö 48. [SY48/2006 Biodiversity economics of the Finnish government] Nature. 142.

54. Newmark, W, D Manyanza, D-G Gamassa, and H Sariko. 1994. The conflict between wildlife and local people living adjacent to protected areas in Tanzania: human density as a predictor source. Conservation Biology 8 (1): 249–255.

55. Nieminen, M. 2008. Luonnonsuojelualueiden merkitys ja käyttö Suomen poronhoidossa Loppuraportti 2008. [Importance of protected areas and the use of Finnish reindeer herding – Final Report, 2008] Porontutkimusasema, Kaamanen: Riista – ja kalatalouden tutkimuslaitos [Game and Fisheries Research Institute, Reindeer Research Station, Kaamanen].

56. Norgaard, R. 2010. Ecosystem services: From eye-opening metaphor to complexity blinder. Ecological Economics 69: 1219–1227.

57. Nuttall, M. 2012. Tipping points and the human world: living with change and thinking about the future. Ambio 41 (1): 96–105

58. Ollila, T. 2008. Maakotkaraportti 2007. Metsähallitus. [Report on Golden Eagles 2007].http://wwwz.metsa.fi/sivustot/metsa/SiteAttachments/Maakotkaraportti2007.pdf . Accessed 12 March 2012. (Unofficial translation)

59. Paanajärvi. 2010. The Paanajärvi National Park. http://parks.karelia.ru/paanajarvi/ . Accessed 2 March 2010

60. Paloniemi, R, and V Vilja. 2009. Changing ecological and cultural states and preferences of nature conservation policy: The case of nature values trade in south-western Finland. Journal of Rural Studies 25 (1): 87–97

61. PAN Parks. 2008. Verification Manual. Principles and criteria 1–3, 4, and 5 January 2008. Endorsed by the Supervisory Board. http://www.panparks.org/learn/apply_for_verification/principles_and_criteria Accessed June 11 2010

62. PAN Parks. 2010. PAN Parks concept, case studies and certified national parks http://www.panparks.org/ . Accessed 24 February 2010

63. Pejchar, L, P Morgan, M Caldwell, C Palmer, and G Daily. 2007. Evaluating the potential for conservation development: biophysical, economic, and institutional perspectives. Conservation Biology 21 (1): 69–78

64. Poropäivät. 2005. Kooste Poropäivien esitelmätiivistelmistä Kaamanen 28.-29.4.2005 [Abstracts of presentations in reindeer husbandry days 28–29 April 2005 in Kaamanen]. Riistantutkimuksen tiedote 199: 1–35. Helsinki, 28 April 2005. (Unofficial translation)

65. Rantamäki-Lahtinen, L. 2008. Porotalouden taloudelliset menestystekijät. [Reindeer husbandry economic success factors] MTT selvityksiä 156. [MTT Research Studies] Agrifood. MTT Agrifood Research Finland, Finland.

66. Raskin, P, F Monks, T Ribeiro, D van Vuuren, and M Zurek. 2005. Global scenarios in historical perspective. In Ecosystems and human wellbeing: scenarios. Volume 2: findings of the scenarios working group of the Millennium Ecosystem Assessment, ed. Carpenter SR, Pingali PL, Bennett EM, BZurek M, 35–44. Island Press, Washington.

67. Reindeer Herder's Association. 2010. Reindeer Herder's Association. Annual statistics. Rovaniemi. http://www.paliskunnat.fi . Accessed 9 March 2010

68. Ruuttula-Vasari, A, and S-K Juvonen. 2006. Oulanka – kuohujen keskeltä kansallispuistoksi. [Oulankajoki: the foam in the middle as a national park] Metsähallitus: Pohjanmaan luontopalvelut. Ostrobothnia Natural Heritage Services, Metsähallitus

69. Rytteri, T, and R Puhakka. 2009. Formation of Finland's national parks as a political issue. Ethics, Place and Environment 12 (1): 91–106

70. Sandell, K. 2005. Access, tourism and democracy: a conceptual framework and the non-establishment of a proposed national park in Sweden. Scandinavian Journal of Hospitality and Tourism 5 (1): 63–75

71. Sandström, C, J Pelllikka, O Ratamäki, and A Sandec. 2009. Management of large carnivores in Fennoscandia: new patterns of regional participation. Human Dimensions of Wildlife: An International Journal 14 (1): 37–50

72. Selby, A, and L Petäjistö. 2008. Residents' adjustments to adjacent national parks: a place-bound structuration perspective. Working Papers of the Finnish Forest Research Institute 90. http://www.metla.fi/julkaisut/workingpapers/2008/mwp090.htm Accessed 15 September 2008

73. Sarkki, S. 2011. The site strikes back: participation and forest governance in northern Finland. PhD thesis. Thule Institute and Discipline of Anthropology, University of Oulu, Finland.

74. Sarkki, S, HI Heikkinen, and R Puhakka. In press. Boundary organizations between conservation and development: Insights from Oulanka national park. World Review of Entrepreneurship, Management and Sustainable Development, Finland.

75. Sarkki, S, and AR Rönkä. 2012. Neoliberalisations in Finnish forestry. Forest Policy and Economics 15: 152–159.

76. Sippola, A-L, H Norberg, M Renko, K Suopajärvi, and T Sutinen. 2005. Petovahinkojen sosioekonominen merkitys porotaloudelle Suomessa – loppuraportti. [Socio-economic importance of predator damages for reindeer husbandry in Finland: Final Report] University of Lapland: Arctic Centre Reports 2005: 44. (Unofficial translation).

77. Skogen, K, I Mauz, and O Krange. 2008. Cry wolf! Narratives of wolf recovery in France and Norway. Rural Sociology 73 (1): 105–133. Publisher Full Text

78. Spence, MD. 1999. Dispossessing the wilderness: Indian removal and the making of the national parks. Oxford University Press, Cary.

79. Sommerville, M, J Jones, and E Milner-Gulland. 2009. A revised conceptual framework for payments for environmental services. Ecology and Society 14 (2): 34.

80. Stenger, A, P Harou, and S Navrud. 2009. Valuing environmental goods and services derived from the forests. Journal of Forest and Economics 15 (1–2): 1–14.

81. Stritih, J, F Mari, and H Karjalainen. 2007. PAN Parks 2007: PAN Parks verification. Oulanka National Park, Finland. Renewal verification report. Principles and Criteria 1–3 and 4–5. PAN Parks.

82. Swallow, B, M Kallesoe, U Iftikhar, M van Noordwijk, C Bracer, S Scherr, K Raju, S Poats, A Kumar, O Duraiappah, B Chieng, H Mallee, and R Rumley. 2009. Compensation and rewards for environmental services in the developing world: framing pan-tropical analysis and comparison. Ecology and Society 14 (2): 26.

83. Thompson, J. 1993. Addressing the human dimensions of wolf reintroduction: an example using estimates of livestock depredation and costs of compensation. Society and Natural Resources 6: 165–179

84. United Nations' Environmental Program (UNEP). 2007. Global Environment Outlook 4. Environment for development. UNEP, Nairobi. http://www.unep.org/geo/geo4/ . Accessed 15 February 2010

85. WWF. 2010. Science: payment for ecosystem services. http://www.worldwildlife.org/science/projects/ecosystemserv/item1987.html?linklocation=topnavdropdownmenu . Accessed 15 February 2010

86. Ylimaunu, T. 2009. Matkailun ympäristösertifiointien tunnettuus ja merkitys luontomatkailijoille: esimerkkinä Oulangan kansallispuisto ja PAN Parks — sertifikaatti. [Environmental certification of tourism awareness and the importance of nature tourists: the example of Oulanka and PAN Parks certificate] Master of Science thesis. University of Oulu, Finland.

87. Zinke, A, G Miller, J Stritih, and P Borg. 2002. PAN Parks 2002: Verification of Oulanka National Park. Verification report. PAN Parks, Finland.

Citations

CHAPTER 1

Baracho MS, Camargo GA, Lima AMC, Mentem JF;Moura DJ, and Moreira J;Nääs IA, Variables Impacting Poultry Meat Quality from Production to Pre-Slaughter: a Review, doi.org/10.1590/S1516-635X2006000400001.

CHAPTER 2

Irais Sánchez-Ortega, Blanca E. García-Almendárez, Eva María Santos-López, Aldo Amaro-Reyes, J. Eleazar Barboza-Corona, and Carlos Regalado, "Antimicrobial Edible Films and Coatings for Meat and Meat Products Preservation," The Scientific World Journal, vol. 2014, Article ID 248935, 18 pages, 2014. doi:10.1155/2014/248935.

CHAPTER 3

Subhas C. Shit and Pathik M. Shah, "Edible Polymers: Challenges and Opportunities," Journal of Polymers, vol. 2014, Article ID 427259, 13 pages, 2014. doi:10.1155/2014/427259.

CHAPTER 4

Alejandro Cifuentes, "Food Analysis: Present, Future, and Foodomics," ISRN Analytical Chemistry, vol. 2012, Article ID 801607, 16 pages, 2012. doi:10.5402/2012/801607.

CHAPTER 5

Joerg Henning, Than Hla, and Joanne Meers, Interdisciplinary Communication of Infectious Disease Research – Translating Complex Epidemiological Findings into Understandable Messages for Village Chicken Farmers in Myanmar, doi:10.1186/2193-1801-3-726.

CHAPTER 6

Katherine M Homewood*, Pippa Chenevix Trench, and Daniel Brockington, Pastoralist livelihoods and wildlife revenues in East Africa: a case for coexistence, doi: 10.1186/2041-7136-2-19.

CHAPTER 7

Hannu I Heikkinen, Simo Sarkki, and Mark Nuttall, Users or Producers of Ecosystem Services? a Scenario Exercise for Integrating Conservation and Reindeer Herding in Northeast Finland, doi:10.1186/2041-7136-2-11.

V

W